T0325678

Biochemical
ECOTOXICOLOGY
Principles and Methods

Biochemical
ECOTOXICOLOGY

Principles and Methods

FRANÇOIS GAGNÉ
Senior Research Scientist
Biochemical Ecotoxicology at Environment Canada
Québec, Canada

With Additional Contributions From
CHANTALE ANDRÉ
JOËLLE AUCLAIR
ÉMILIE LACAZE
BRIAN QUINN

AMSTERDAM • BOSTON • HEIDELBERG • LONDON
NEW YORK • OXFORD • PARIS • SAN DIEGO
SAN FRANCISCO • SINGAPORE • SYDNEY • TOKYO
Academic Press is an imprint of Elsevier

ELSEVIER

Academic Press is an imprint of Elsevier
32 Jamestown Road, London NW1 7BY, UK
225 Wyman Street, Waltham, MA 02451, USA
525 B Street, Suite 1800, San Diego, CA 92101-4495, USA

British Library Cataloguing-in-Publication Data
A catalogue record for this book is available from the British Library

Library of Congress Cataloging-in-Publication Data
A catalog record for this book is available from the Library of Congress

ISBN : 978-0-12-411604-7

For information on all Academic Press publications
visit our website at elsevierdirect.com

Printed and bound in United States of America

14 15 16 17   10 9 8 7 6 5 4 3 2 1

**Working together
to grow libraries in
developing countries**

www.elsevier.com • www.bookaid.org

CONTENTS

LIST OF CONTRIBUTORS

Chantale André
Emerging methods, Aquatic Contaminants Research Division, Environment Canada, Montréal, Québec, Canada

Joëlle Auclair
Emerging methods, Aquatic Contaminants Research Division, Environment Canada, Montréal, Québec, Canada

Sylvie Bony
Ecole Nationale des Travaux Publics de l'Etat France

Alain Devaux
Ecole Nationale des Travaux Publics de l'Etat France

François Gagné
Emerging methods, Aquatic Contaminants Research Division, Environment Canada, Montréal, Québec, Canada

Émilie Lacaze
Institut Armand-Frappier-INRS, Laval, Québec, Canada

Brian Quinn
University of West Scotland, Scotland, UK

DEDICATION

I dedicate this book to Lucie, my life-long companion, for her unconditional support in the many projects I undertake. I also dedicate this book to my daughters Geneviève, Catherine and Valérie for providing me with reasons to fight for the protection of the environment and to my parents for their support during my studies in biochemistry and thereafter. I also dedicate this book to my mentors Prof Francine Denizeau (in memoriam) and Dr Christian Blaise for their guidance, foresights and for opening my eyes to the science of ecotoxicology.

François Gagné
January 2014

The science of ecotoxicology is the study of contaminant exposure and its effects on all organisms in the biosphere, including humans. This discipline started in the late 1960s and was defined as the study of the contamination of the biosphere and the resulting toxic effects to all life forms. The scope of ecotoxicology is quite broad because it deals with the studies of the fate and effects of contaminants from the molecular level to the ecosystem. To address this complexity, ecotoxicology was subdivided into three phases or pillars: exposure and bioavailability of contaminants, toxicity evaluation of substances with their mixtures, and biomarkers to determine the long-term risk of pollution in organisms and corresponding populations. This relatively new science was further complicated by the observation that not only chemical agents are at play but the presence of physical (radiation, temperature variations) and biological agents (infectious diseases) are also involved. Hence, there is a need to understand the cumulative contribution of these agents to organism health and the maintenance of populations. In addition, the advent of new exotic products such as nanotechnology, climate change, and landscape transformation will bring new types of toxic interactions that could overwhelm the ability of organisms to cope with existing environmental stressors such as prey—predation changes, eutrophication, food availability, water levels (and quality), and seasonal temperature changes.

Biomarkers are broadly defined as a measured variable that results in or follows the interactions of (toxic) agents with a molecular or physiological target. It is generally recognized that fundamental interactions occur at the biochemical level in cells provided that this interaction forms the basis of initiating toxic events in organisms. Some biomarkers are more specific (more closely related) to the toxic agents so they can be used in testing schemes to identify the contribution or the cumulative effects of various toxic agents. These types of biomarkers are particularly valuable since organisms are seldom exposed to one compound or toxic agent at a time. Indeed, organisms are usually exposed to low concentrations of hundreds and perhaps thousands of chemicals in their lifetime. Biochemical markers could also determine some biochemical changes in organisms that could lead to deleterious effects on organism health in the long term. To provide a clear representation for the reader, the occurrence of high lipids (cholesterol) and oxidative stress (inflammation) in the plasma is associated with the degeneration of blood vessels and related diseases in humans. Hence, the measure of plasma lipids or cholesterol represents a biomarker of risk in developing a vascular pathology in time. In other words, a biomarker could represent a biochemical condition that could lead to damage at a higher level of biological organization. In order to predict effects at the level of organism's health and maintenance of population, we need to understand how a biochemical change resonates at a higher level of biological organization. Biochemical markers could contribute to this understanding and find their use and place in ecotoxicology.

When I started as an experimental scientist, many techniques were not always readily available in the laboratory. We (my colleagues and myself) often had to develop and adapt methods from those reported in the primary literature. We also had to examine hypotheses on issues related to a particular biochemical effect of given pollutants to a physiological target in organisms under budget and time constraints. As a preliminary approach we sometimes first used generic and cost-effective biomarkers to determine whether a type of effect occurred in exposed organisms before using more sophisticated and expensive techniques or approaches. We quickly came to realize that these preliminary testing tools were of great interest to laboratories with tight budgets (which is often the case for most laboratories) or located in remote areas. We discovered that not only highly sophisticated techniques are valuable in ecotoxicology but generic ones as well. Another advantage of generic assays, which are based on a particular and fundamental property of the biomarker, is their potential adaptability to be used in new species. These experiences made me decide to produce a book of methods in biochemical ecotoxicology that illustrates both generic and highly sophisticated and species-specific biomarkers that are accessible to many laboratories with budget and other constraints. We also realized that biochemical markers should stand the test of time, be cost-effective, simple to use, and accessible by science laboratory students without major investment. In this respect, this book will assist the experimenter in ecotoxicological sciences to quickly start biochemical-based assays on various organisms. Our laboratory receives many requests by graduate students, young scientists beginning in this field, and other researchers getting acquainted with a new area of research to learn these techniques. Each chapter of this book could be considered comprehensive on its own and follows a common structure to enable the reader to quickly learn and try the assay. Each chapter has a brief introduction showing the context and principle of the assays, a section on reagents and solutions required for the procedure, data handling, and references. In some instances, case studies or examples are provided to demonstrate how the methods can be used. I am confident that this book will be well appreciated by both researchers and students who need quick and direct guidance in performing biochemical assays in ecotoxicology.

François Gagné
Montréal, January 2014

ACKNOWLEDGMENTS

The principal author is very grateful to the contributions of Dr. Brian Quinn for the preparation and use of cell cultures in toxicity investigations in Chapter 3, Dr. Émilie Lacaze, Sylvie Bony and Alain Devaux for the contribution on the COMET assay in Chapter 10, Chantale André for her contribution on gene expression assays using quantitative reverse-transcriptase polymerase chain reaction in Chapter 4, and to Joëlle Auclair for her contribution on protein transfer and detection techniques (Western blots) in Chapter 5 for the detection of protein-based markers.

François Gagné

Biochemical ecotoxicology is the study of the effects of contaminants in ecosystems using biochemical methods. The scope of this discipline is vast and complex because it implicates exposure to low concentrations of mixtures in the long term (chronic), encompassing most of the life stages of organisms. Although the toxicity of chemicals mostly depends on the physicochemical properties of chemicals, the resulting effects will depend on the lifestyle of the organisms, such as nutritional status, reproduction activity, prey—predation pressures, and habitat characteristics. The advent of climate change adds another level of complexity to predicting the toxic outcomes of pollutants, which is a challenge for the risk evaluation community. The biochemical approaches in ecotoxicology have the advantage of finding early warning signals at the molecular level, which can lead to effects at higher levels of biological organization. The interactions of chemicals with biochemical targets in cells or tissues that could lead to altered function represent a special type of "biomarker" in toxicology. There are many definitions proposed in the literature for biomarkers, but we think this one best reflects the biochemical markers: the term biomarker is used in a broad sense to include almost any measurement reflecting an interaction between a biological system and an environmental agent, which may be of chemical, physical, or biological origin [1].

This book presents practical methods and approaches used in the field of biochemical ecotoxicology. It is destined for laboratory investigators involved in environmental toxicology investigations for the protection of wildlife and aquatic ecosystems against pollution and other stressors such as climate changes. As explained previously, biomarkers are defined as measures highlighting an interaction between a xenobiotic and other stressors (e.g., temperature) and a biological target in the physiological sense [2]. Biomarkers are also determined at different levels of biological organization starting at the molecular, subcellular (organites), cellular, organ/tissues, organism, and population levels (Figure 1). Often called the ecotoxicology continuum, biochemical ecotoxicology is mainly focused at the molecular and subcellular levels and the conditions or criteria by which a molecular toxic interaction resonates at a higher level of biological organization. For example, the expression of genes involved in the biotransformation of xenobiotics that leads to increased activity of a given enzyme, which in turn leads to potentially toxic protein or DNA adducts. Changes at the molecular level are dynamic and occur rapidly in low timescales (seconds to minutes and up to many days), while effects at the tissues/individual level are manifested generally at later times (hours and days to years). In some cases, biochemical alterations take place at the early stages of

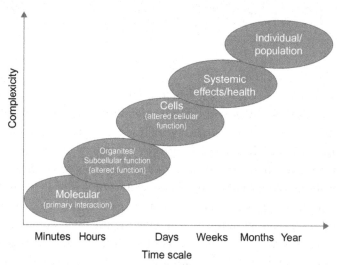

Figure 1 Levels of biological organization in the manifestation of toxicity.

pathogenesis, but the observed effects are not always irreversible, hence their usefulness in prevention. In other cases, responses observed at the biochemical level (i.e., gene expression changes) do not imply an impact on the organism's health status and performance, hence their lower "physiological or ecological" relevance. Endless arguments about which levels are more physiologically or ecologically relevant for ecotoxicology studies are considered useless and outdated. It depends on the questions being asked in a given study and the context of pollution studies. The science of ecotoxicology is hierarchized, i.e., one compartment is a part of another compartment at a higher level of biological organization. It is generally accepted that an impact at any level is likely to produce changes at a higher level provided the intensity of the responses and/or duration of effects are sufficient. It is a question of knowing when an alteration at one level (e.g., enzyme inhibition) will translate to an effect at the next level of complexity (e.g., altered metabolic pathway from an inhibited enzyme). From the risk assessment perspective, biomarkers that provide information on health status toward survival, growth, and reproduction are particularly relevant for the protection of population toward pollution and other stressors involved in these times of climate change.

Investigators in ecotoxicology are often confronted to examine or work with all types of species from which no genetic information or specific antibodies or other probes are (yet) available. They have to fall back on generic tests, which focus on special biochemical properties of the biological target. Indeed, modern and state-of-the-art techniques are highly dependent on the information of the test species under investigation (availability of specific gene sequences or antibodies). In the absence of such information with new species, the investigator has to rely on "generic" biomarkers, which take into consideration a particular property of the analyte, e.g., high calcium and phosphate content (vitellogenin), heat-stable metal binding proteins (metallothioneins), or the capacity to catalyze a given chemical reaction (enzymes) [3].

Although these methods are sometimes less specific and not as sensitive as immunoassays or quantitative PCR techniques, generic tests are an ideal choice for screening and exploration studies because they are more accessible (cheap, quick, and require only basic equipment) compared to the more sophisticated modern assays. These tests "stand the test of time" because of their ability to be used on various species, in different laboratories, and under budget constraints. For example, the acetylcholinesterase assay based on the dithionitrobenzoate (DTNB) thiol reactive reagent has been actively used since 1961 [4]. Indeed, not all laboratories have the financial capacity or infrastructure to purchase highly specialized equipment. Sometimes an investigator needs to explore whether an ecotoxicological problem occurs at a given site without making a major investment. For example, the evaluation of the estrogenicity of a municipal effluent outfall in a river could be easily verified by measuring vitellogenin (egg yolk protein, which is regulated by estradiol-17β) in the plasma of fish or the gonad tissues in mussels using a generic assay based on the alkali–labile method [2,5]. This assay will require only basic equipment (centrifuge and spectrophotometer) and is based on a cheap and accessible inorganic phosphate assay. Hence, the use of cheap and rapid methods to explore a specific topic or examine a hypothesis could be valuable when human and financial resources are limited. This is especially useful for remote laboratories or laboratories with limited financial capacity. That is not to say that if resources are available, high-throughput screening tools by DNA microarrays or proteomics, for example, could be chosen for biomarker mining and discovery.

This book offers a balance between generic biochemical techniques and highly specific high-throughput techniques such as quantitative polymerase chain reaction and enzyme-linked immunoassays. We think that this book will introduce novice and more experienced scientists to techniques that are accessible to all laboratory sizes and budgets. This work will provide a way for investigators to quickly start biochemical assays in an ecotoxicology laboratory. Guidelines in the application of biochemical markers in research and monitoring studies are also provided that will give insights on quality control and assurance aspects of studies using biomarkers. This book was created with these principles in mind to enable investigators to quickly start biochemical ecotoxicity assessments in their respective laboratories. It is destined to be found on both the laboratory bench and office desk. Each chapter begins with a brief introduction to set the context of the proposed analyses with a viewpoint on the relevant biomarkers. The general structure of the chapters will follow a simple, direct, and comprehensive layout: introduction and principle, reagents and materials, procedure, and data calculation. In some cases, some chapters will end with a case study to provide examples. We are confident that this book on principles and methods of biochemical ecotoxicology will help initiate real-life experiments in biochemical ecotoxicology.

François Gagné, Environment Canada
Chantale André, Environment Canada
Joëlle Auclair, Environment Canada
Émilie Lacaze, INRS-Institut Armand Frappier
Brian Quinn, University of the West Scotland

REFERENCES

[1] Benford D, Hanley AB, Bottrill K, Oehlschager S, Balls M, et al. Biomarkers as predictive tools in toxicity testing. ATLA 1999;28:119−31.
[2] Gagné F, Blaise C. Review of biomarkers and new techniques for in situ aquatic studies with bivalves. Environmental toxicity testing. New York: Blackwell Publishing; 2005, 205−228.
[3] Gagné F, André C, Blaise C. Increased vitellogenin gene expression in the mussel *Elliptio complanata* exposed to estradiol. Fresenius Environ Bull 2005;14:861−6.
[4] Ellman GL, Courtney KD, Andres Jr V, Featherstone RM. A new and rapid colorimetric determination of acetylcholinesterase activity. Biochem Pharmacol 1961;961:88−95.
[5] Verslycke T, Vanderbergh GF, Versonnen B, Arijs K, Janssen CR. Induction of vitellogenesis in 17alpha-ethinylestradiol-exposed rainbow trout (*Oncorhynchus mykiss*): a method comparison. Comp Biochem Physiol 2002;132:483−92.

CHAPTER 1

Quantitative Assessments of Biochemical Analyses

François Gagné

Chapter Outline

1.1 GENERAL OVERVIEW

In ecotoxicological studies, the early biological (toxic) effects of chemicals are determined by the use of biochemical markers. Biomarkers rely on chemical analyses of biological macromolecules such as deoxyribonucleic acids, polypeptides/proteins, and membrane assemblages. The ability of proteins to catalyze biochemical reactions (enzymes) is noteworthy and constitutes an important part of quantitative biochemistry. The chemical analyses are identical to quantitative analysis of chemicals in the classic sense (i.e., evaluation of mercury by cold vapor atomic absorption spectrometry) but differ because the analysis is done in relation to a given biological function that often involves highly complex molecules (i.e., mercury binding thiol-rich proteins such as metallothioneins, MTs). As another example, the quantitative evaluation of serotonin, a

Biochemical Ecotoxicology
DOI: http://dx.doi.org/10.1016/B978-0-12-411604-7.00001-5

derivative of the amino acid tryptophan, by liquid chromatography with electrochemical detection is an example of chemical analysis, but the interaction of serotonin on membrane receptors in nerve cell membrane is biochemical because it intervenes in complex biochemical interactions with macromolecules. These assays are practiced in the context of hypothesis testing, i.e., in studies to determine negative interactions, if any, of exposure to a given xenobiotic to a physiological function in the exposed organisms or cells. Moreover, test organisms in the real world are genetically heterogeneous and under various habitat pressures such as nutritional stress, temperature, climate variations, and reproductive cycle. This brings more interindividual variability in the measures [1]. This variation is added to the variation of the method of biochemical analysis. It is therefore essential to determine both the variation induced by the method of analysis and interindividual variation if one wants to highlight toxicological effects. Hence, in the effort to better understand the variability of biological responses, we must acquire a good understanding on the life cycle of the organism and seek the normal range of responses of the biomarker over time (seasonal variability) in a given age range [2]. However, the knowledge of all these aspects might be difficult when dealing with a new species not before examined. The understanding of the "normal" range of responses of biomarkers actually represents an important aspect of ecotoxicology research and monitoring.

Studies in biochemical ecotoxicology bring about new methodologies, new versions, or adaptations of existing methods. In most cases, the methods were developed for mammals where modifications and adaptations are made when applied to a new species. Moreover, there is a big step between a newly developed methodology for a given research study and for a method for monitoring purposes. Many biomarkers are often measured by many means (methods), each having their advantages and caveats. For example, methods developed for a given research project require different types of validation than monitoring types of studies. Some of these aspects will be discussed in this chapter. For example, the family of heavy metal binding proteins, including MTs, is involved in the sequestration and protection of toxic metals and they are determined by many methodologies such as the silver or cadmium or mercury saturation assays, pulse polarography, enzyme-linked immunoassays, by mRNA determination using quantitative polymerase chain reaction (qPCR), high-pressure liquid chromatography (HPLC), and gel electrophoresis. Each of these methodologies has advantages and caveats in respect to rapidity, cost-effectiveness, sensitivity, specificity, and reproducibility. It is needless to argue that only one methodology should be used for a given project. Many of these assays are partially quantitative where the relative levels of the biomarker are determined in contrast to the absolute amount. In addition, most biomarkers require normalization against biomass, which is also determined by a variety of approaches. For example, the biomarker could be normalized against wet or dry tissue weights or total protein or DNA contents. For mRNA determinations by qPCR, the levels of mRNA are usually normalized against total mRNA levels or with the levels of housekeeping genes provided they are stable in the conditions of the experiment (which is often not the case). These aspects will be discussed in detail in Chapter 4. Hence, biomarkers are not only determined by numerous methodologies, but their normalization could be determined by a variety of methods as well. This introduces another level of complexity for quantitative assessment of biomarkers between laboratories and projects. From a monitoring perspective, the use of certified reference material for biomarkers to determine the precision and trueness of the method is lacking in most cases. These materials are used to validate the various laboratories to produce similar results, especially in the context of long-term temporal and spatial surveys. Nevertheless, some precautions or steps could be undertaken to improve the reproducibility and reliability of quantitative biomarker assessments in a research setting [3]. This chapter consists of guidelines to assist the laboratory investigator (or research assistants, students, etc.) in generating sound bioanalytical data and in establishing the context and limits of the analyses. As discussed earlier, the MT biomarker will be used as an example model for clarity purposes for the reader. The scope of this chapter is not only to provide quality control and assurance notions in a strict sense but to adapt them in a research context in biochemical ecotoxicology studies. The principle of quality control and assurance for standardization are similar but not identical for standardization methods in monitoring, which makes use of certified reference material [4]. Notwithstanding, good laboratory practices should always be practiced in the most rigorous fashion possible.

In a research laboratory, various steps are taken to secure and validate the data when developing a methodology and generating data (Table 1.1). These criteria will be explained in this chapter. First, a laboratory notebook should accompany the experimenter at all times where all relevant information is duly noted on a

Table 1.1 Levels of Control for the Generation of Biomarker Data

Control Level	Parameters		Comments
1. Type of project	Objectives Hypothesis Protocols	Replication Standard solutions Instrument readings Secure rough data	To be included in the laboratory notebook
2. Replication	Analysis Experiment		Replication used Determine the repeatability of the experimental design
3. Variability	Method of analysis Biological variability		Determine the variation of the method and the biological variation (interindividual)
4. Validation	Correspondence between the biomarker and means of normalization	Determine the constant slope between the biomarker and biomass evaluation	Find the linear range of the rate of change of the biomarker with the rate of change in the biomass
5. Reproducibility	Calibration (standardization) Determine the stability of the signal of a standard corrected against a blank Stability of a reference material at each time of analysis of the biological samples (composite sample)	External standard Internal standard (matrix effects) Positive control (with a known inducer) Quality control chart of the standard signal in time Confirm the stability of the biomarker signal over time using different reagents/standards preparations and even with different analyst by including the analysis of the same sample at each sample batch	Induction of the biomarker with a known and previously reported inducer of the biomarker Determine the method's trueness in time and with different analyst, instruments, and reagents A mean to validate that the method used was always correct in quantifying the biomarker in time
6. Data archives	Store all data in a safe place: rough and compiled data should be included	Hard drives Networks	Secure data for long-term use

day by day basis. Moreover, bioanalytical data should be protected and well identified in the storage media. It is important to safely store the data and identify the
quality control measures from which the data were generated: the protocol, the
composition, and date of preparation of blanks, standards, type of samples, and
replication of the analysis. First, archives call for laboratory notebooks where all
the methodological and procedural details including any small modifications, if
present, are found. Replication and the method used for calibration/standardiza-
tion should always be noted in the log book for the production of any (semi-)
quantitative bioanalytical data. Details on the experimental design used and the
variation of the methodology with its limit of detection should be included as
well. In a research laboratory, it is possible that specific standards are lacking. In
this situation, recommendations are proposed to provide some extent of validation for the method of analysis. Finally, elements of quality control proposed in
situations were certified reference material for biomarkers in a given test species.
These elements will permit the undertaking of spatial or temporal surveys in
ecotoxicological investigations.

1.2 REPLICATION IN DATA ANALYSIS

In all research activities, replication is required at both the analysis and experimental steps. Each blank, standard (if available), and sample should be analyzed
using three replicates with the instrument. In some instances, when resources are
limited or the method was proved to be very reliable (like a pH or weight measurements), the samples could be analyzed once, but one test sample in the lot is
analyzed four to eight times to have at least one good estimate of the method's
variability. The number of replicates depends on the intrinsic variation of the
methodology, type of instrument, and the biological variability. Replication at
the analysis step permits the definition of the variation of the analytical method
or the precision. The experimental precision consists of the dispersion of the data
around a central attractor such as the mean or median. Precision is usually
expressed as the coefficient of variation of the mean ((standard deviation/
mean) $\times 100$) for normally distributed data. To obtain good estimates of precision, the sample must be analyzed several times (ideally four to eight times,
depending on the intrinsic variation of the assay) at least once during the analysis
batch. The number of samples should take into account the number of individuals involved in a given study in an attempt to determine the sensitivity of the

assay since the biological responses should produce changes that are above the variation of the method. In other words, the *intra*individual variation (variation of the biomarker during analysis) should be lower than the *inter*individual variation to obtain a significant effect between groups of individuals.

At the level of biomarker data, care should be taken during the sequence of analysis of the blank, standards, and samples by the instrument to control for any drift in the performance of the instrument. With microplate readers, readings are taken within seconds depending on the technology, thus the variation of measurement within the microplate is expected to be negligible. This is not necessary for absorbance readings given the changes in absorbance are obtained by the ratio in the light intensity of the instrument in the presence (Is) and absence (Io) of the sample (log Is/Io as defined by Beer—Lambert's law). If the readings are repeated after 30 or 60 min, the rough data could differ between reading times because the performance of the instrument could vary in time (e.g., the lamp energy could vary in times and have major effects on fluorescence readings or change in a photomultiplier tube efficiency for luminescence readings). Hence, there is a need to include blanks and standards for each sequence analysis. Some instruments are equipped with internal reference, which directs correction of the data after each reading by the software.

Variations in the efficiency of an instrument are more important with instruments that require samples to be determined over minutes to hours of analysis time. For example, if the performance of the instrument changes by 25%, then this could introduce important variability in the data. In these cases, it is recommended to measure the blank and standards at the beginning of the analysis, measure 5—10 samples, and measure again the blank and one standard and so forth. With this approach, one could control for any drifts of the instrument performances. The instrument's reading efficacy depends on the type of instrument (HPLC columns and detectors, fluorescence or luminescence readers) and is characterized as change in (sample signal—blank signal)/concentration of the standard sample. This metric is important if measures are made on different instruments across laboratories. Indeed, the efficacy of the instrument changes in time and with different instruments.

1.3 REPLICATION OF THE EXPERIMENT

The notion of replication should be extended at the level of the experiment itself, especially in a research context. This approach differs from the context of routine analysis where a sample is analyzed and compared with a standard and certified reference material in a similar matrix. Take a study consisting of the evaluation of MT in mussel populations at two sites (one contaminated by metals and one pristine). To start, eight individuals were collected for the determination of MT in the digestive gland (equivalent to the liver) at each site. The MT assay was practiced in duplicate for each individual. In theory, if the method is reproducible and precise, on the order of 5% or less, then the required number of individuals would depend on the interindividual variability (biological variation). If the interindividual variability is at 30% (coefficient of variation of 30%), then the mean level of MT should be higher than the mean + 30% value to detect a difference. If the sample size is sufficient to appraise the state of the population (we assume for simplicity that the gender, size, and age of sampled mussels are constant) then the repetition of this experiment should provide similar results. If not then the $N = 8$ individuals was not representative of the local population and more individuals should be determined. Usually the experiment is repeated three times to definitely ensure the replication number was enough. In other words, we have to make sure the differences in MT levels between the two sites are not the result of method variation or insufficient sample size. We must determine the number of replicates for analysis (which depends on the methodology used) and the number of individuals (sample size), which depends on the biological variation. The latter case is a question of power analysis, that is, the number of replicates required for a significant effect above a threshold, usually $> 30\%$ of the mean, since the normal biological variation is usually in this order. When the sample size is not sufficient, repeating the experiment will provide confirmation if the observed effects are real or not.

1.3.1 Calibration or Standardization

The methodology used should include analytical blank and standard samples for calibration. In addition, the reading of the test sample (unknown sample) should be within the readings of the blank and standards where a linear relationship exists. For example, if the absorbances of the blank and highest standard are 0.01 and 0.22, respectively, then the sample absorbance should be included within this interval. The blank is defined as the matrix in which the analysis is performed without the biomarker (or very low levels depending on the type of the assay) of interest. Standardization of the method supports two roles: (1) quantitative determinations, and (2) the establishment of the limit of detection of the methodology. The limit of detection is defined two ways:

theoretical and operational. The theoretical limit of detection of the biomarker is the value corresponding on the mean signal $+2 \times SD$ of the blank. For example, if the blank samples give a mean of 0.1 ± 0.02 then the theoretical detection limit is $0.1 + (2 \times 0.02) = 0.14$. Hence, there is a need to have replication during analysis. The operational limit of detection corresponds to the lowest measured standard value. This definition is more conservative and depends on the linear relationship between the analytical signal and the concentration of standard used. Since these definitions are different, it is important to identify which one is used.

1.3.2 Using Specific and Generic Standards

When a specific standard for a given species is lacking, it is possible to use a standard from another species provided the biomarker shares similar properties in each species. For example, rabbit MT standards could be used to standardize the assay dealing with other species such as fish, mussels, or rats. The standard addition approach, where the standard is directly added to the unknown samples, could be used to account for unsuspected matrix effects that could arise in samples from other species. In this case, these data could be expressed in an equivalent amount of added standard. For the MT example, MT levels in the digestive gland of mussels using rabbit MT would be expressed as nanomoles of rabbit MT equivalent/biomass of the digestive gland.

Standards can be added externally in a separate tube without the test sample or directly in the test sample. When the method is impervious to interferences from the complex biological matrix, an external calibration curve could be produced where pure standards are serially diluted in the assay buffer. In this situation the data point starts at the origin (Figure 1.1A) and the presence of sample does not contribute to the signal or is effectively removed by an analytical blank. The concentration of the test sample is then calculated by the linear regression line: Analytical Signal (y) = value (b) + Slope $(a) \times$ Standard concentration $(x) \rightarrow x = (y - b)/a$. This procedure holds if the presence of the biological matrix does not interfere in any way with the analytical signal y.

When the biological matrix could produce interfering effects (i.e., decreasing or increasing the analytical signal), the standard curve is constructed in the presence of the test sample (Figure 1.1B). In this situation the absence of the added standard yields a signal (interference) and changes in the analytical signal output (the slope is either lower or higher). In this situation, the complexity of the matrix introduces a

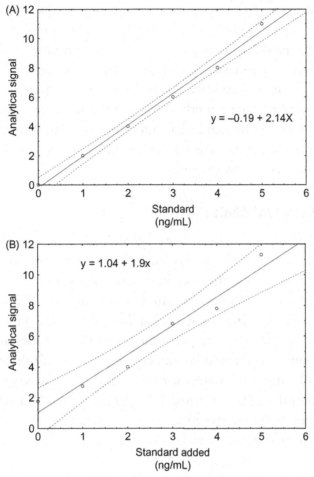

Figure 1.1 Typical calibration curve between the analytical signal and the concentration of the standard. An external calibration curve is shown in A and an internal calibration curve using the standard addition method in B.

bias toward the analytical signals. In this case the standards are added directly in a constant volume of the test sample. The concentration of the analyte in the test sample is then extrapolated by giving a value of 0 to the analytical signal (y), for example: $0 = 1.04 + 1.9 \times$ (standard concentration) $\rightarrow$ standard concentration (x) $= |(-1.04/1.9)|$. This procedure, albeit more robust, requires the construction of a standard addition in each test sample, which can become time-consuming. When the volume of the test sample is limited, one standard addition could be added to one of the test samples. The sample concentration is then calculated as follows: concentration of sample = [Analytical Signal (sample − blank)/Analytical Signal (spiked sample − blank)] × added standard concentration. The standard addition method represents a convenient means to control for matrix interference or when this effect in uncertain (e.g., when working with novel organisms). If no interference exists then the standard addition method will produce results similar to the external standard curve; in other words, if the slopes are equal then there is no need to proceed with the standard addition method. This verification could be done at the start of the project when dealing with a new biological matrix such as different organs/tissues or new species.

1.4 REFERENCE SUBSTANCES

When no standard exists for a given species and when no other standard exists from other species, the methodology could be validated by using a reference toxic substance known to induce the biomarker of interest. The ratio between the biomarker level of the control organism and biomarker level of the induced organism should be the same between batches of analysis. This is particularly important when the analyses are performed at different times and in separate batches. For example, MT could be induced in the liver of fish exposed to cadmium or zinc (0.5 mg/L in water) for 96 h at 15°C. The ratio of MT levels between the control and the exposed fish could be used as a reference point. The liver samples could be stored at −85°C as reference samples and they could be tested at each batch of analysis in time.

1.5 DEFINING DETECTION LIMITS IN THE ABSENCE OF ANALYTICAL STANDARDS

As described previously, the operational detection limit is defined as the smallest standard used during calibration or the concentration producing a signal $\geq 2 \times$ the SD of the analytical blank for the theoretical limit of detection. For the latter, a signal above this threshold usually is statistically significant at the $p < 0.05$ level. When analytical standards are available the former method is the preferred procedure. In a research context, standards are sometimes not commercially available and the production of standards is outside the reach of the laboratory. In this situation, we can establish another operational detection limit of the methodology. Let's come back to the MT in the digestive gland example (eight mussels from a reference and metal-contaminated site).

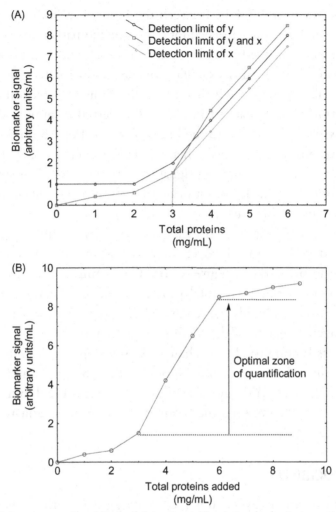

Figure 1.2 Defining an operational limit of detection without standards. An operational (lower) limit of detection could be determined between the biomarker signal and the normalization endpoint, here as total proteins (A). The upper limit of quantification of the biomarker and the normalization method should also be known (B).

We could prepare a representative homogenate extract of the digestive gland and run the assay on serial dilutions of the homogenate. The ratio between MT levels normalized against total proteins should remain constant throughout dilutions and the sample could be used within this boundary. When approaching the limit of detection of either MT or protein assessment methods, the slope between total MT or proteins will change (Figure 1.2A). This imposes a limit of detection of the method that is based on the slope between the biomarker and biomass endpoint. When the limit of detection is reached for either the biomarker or the normalizing assay, the slope between the biomarker and its normalization unit will change, as shown in Figure 1.2A. In this example, the operational detection limit of the methodology will depend on which method the limit of detection reached; for limit of detection x, it is reached first then the detection limit is 3 (total proteins $\times$ axis), which corresponds to the biomarker signal of 1.5 on the y-axis. For the limit of detection y, it is reached when the detection limit is 2 for total proteins (x-axis) corresponding to a biomarker signal of 1 on the y-axis. This approach gives the boundary of the minimal amount of biomass (in protein units in the example) required to measure the biomarker.

At the other side of the limit of detection, the upper limit of quantification should also be determined to discover the maximal amount of biomass that can be tested for the biomarker (Figure 1.2B). For example, the upper boundary would correspond to 6 mg/mL total proteins with a biomarker of 8 arbitrary units. Indeed, using a highly concentrated homogenate extract could overwhelm the reagent capacity of the methodology or introduce new interferences or matrix effects leading to altered changes in the slope of the biomarker with its biomass. Hence, the experimenter should define the upper and lower limits of the assay when standards are absent to determine the biomarker in the region where the slope between the biomarker and biomass unit is invariant, as depicted in Figure 1.2B.

1.6 REPRODUCIBILITY

There is a distinction between the methodological and the experimental or biological reproducibility. The methodological reproducibility is characterized by the capacity of producing constant data regardless of "when and where" the assay is practiced. Indeed, a reproducible method will produce similar data independent of the analyst or technician performing the data, the time and period of the analysis, and the instrument in a given laboratory.

1.6.1 Reproducibility at the Calibration Level

The slope between the analytical signal and the concentration of a standard should be constant regardless of the time and where the analysis was produced. Moreover, the signal difference between the blank and the standard should be relatively constant for a given instrument. These criteria are closely related to the stability of the biomarker and the reagents used to quantify it at the day of analysis. It is important to mark in the laboratory notebook the date of preparation of the reagents and standards, commercial origin, and the lot number when possible.

1.6.2 Reproducibility at the Level of the Experiment

As already mentioned in this chapter, the experiment should be repeated a number of times to confirm whether the produced data lead to similar results when possible, for example, if the ratio of the mean value for MT from the metal-contaminated site/mean value from the reference site is 4 for a given experiment. Repeating this experiment should give a similar ratio, i.e., within the normal variation of the methodology. If the ratio differs markedly (more than the interindividual variability) then this could indicate a problem with the biomarker analysis or the execution of the exposure experiment.

1.7 USING REFERENCE MATERIAL TO STANDARDIZE METHODOLOGIES

It is recommended that all the analyses are made during the same day or analysis batch in order to eliminate temporal drift of the assay methodology. To control for these methodological drifts (related to the method's trueness), a reference material could be produced in-house, which consists of preparing a similar sample and storing separate aliquots at $-85°C$. At the given day of analysis, an aliquot of the reference material is taken for biomarker determinations. At each time of analysis, the reference material should always give the same biomarker value, i.e., within the confidence interval or normal variation of the methodology. The reference material approach consists of producing our own in-house material. In some cases a certified reference material could be produced by a national laboratory using strict procedures, but these have yet to be available for biomarkers in ecotoxicology. For example, if the MT biomarker in the digestive gland is to be examined at 10 sites (8 mussels per site), then a pool of digestive glands (10−50 mussels) could be prepared, stored in separate aliquots, and used as the reference material. Of course, the digestive gland extracts would be prepared using the same conditions such as tissue grinding apparatus, homogenizing steps, temperature, and buffer. The reference material would be used at each day or time of analysis. If it takes one day of analysis time per site, then the project will require 10 separate periods of

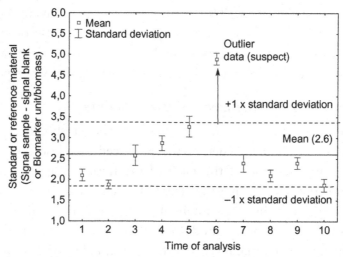

Figure 1.3 Quality control chart illustrating a biomarker from a prepared reference material or a fixed concentration standard determined through 10 days of analysis. These data should be distributed around the mean value (filled line) with SD (dotted lines). Data value outside the SD is considered a suspect outlier and should be either removed from the data set or reanalyzed after making the appropriate revisions of the method.

analysis. For each period of analysis, the reference material is measured (in quadruplicate) where it should give a value within the normal variation of the methodology. The reference material approach is used to track the time at which the measurement was done and determine the trueness of the method in time [5]. This approach enables one to track the performance of the method of analysis in time and in different laboratories. The second option consists of using a standard from another species that is commercially available to ensure the difference between the standard and blank is constant in time. A quality control chart could be produced to identify which samples or day of analysis yield data outside the normal variation of the method. In Figure 1.3, each data point is either a fixed standard corrected against a blank or a normalized biomarker value (biomarker unit/biomass unit) that was analyzed at each period. The SD of the data points corresponds to the methodological variation. The method should always give, in principle, the same value, i.e., a value within the normal variation of the method (within the SD or the 95% confidence interval). The true "normal" variation of the method in Figure 1.3 is depicted as the solid line with the corresponding SD (dotted lines) for all analysis time. At time 6, the obtained data are outside the upper SD of the method and considered an outlier or suspect. Ideally all the samples and standards (blanks) should be reanalyzed after rechecking/re-preparing the reagents or bringing the proper corrections of the methodology. If this is not possible, then the data from all samples analyzed at that time should be removed from the dataset. In some instances (for exploratory purposes), a correction factor could be used to bring down the mean data value at time 6 to the mean value of the method (solid line) for all the samples analyzed at time 6. The correction factor is derived as follows: mean value at suspect time/mean value of the endpoint: $4.9/2.6 = 1.88$. Hence, all the biomarker values obtained at day 6 are divided by 1.88. However, this approach should not be applied as a final measure of the biomarker at this time but only as a preliminary view of the biomarker response after correcting for methodological drift. This is especially important when each time period corresponds to a different treatment group or study site or time survey. Some laboratories take the samples randomly and sometimes blindly (the sample identity is hidden and replaced randomly by numbers) to alleviate potential artifacts and subjectivity.

The use of quality control charts determines the performance of the methodology over different times (and different laboratories). When the production of an in-house reference material is not possible (low availability of biological material or test species), the mid standard value corrected against the blank could be used. This approach is related to the precision of the method at each time of analysis. In most cases, values outside $2\times$ the SD are considered an outlier or aberrant value. In conclusion, when the assays are performed at different times, which involve different preparations of reagents and standards for calibration, the use of a reference material is recommended. This procedure establishes the trueness of the method, which is the normal variation of the methodology applied at different times regardless of the analyst, reagent, and calibration used.

1.8 NORMALIZATION APPROACHES

Most biochemical endpoints (enzyme activity, particular protein, metabolite, etc.) are normalized against a measure of biomass such as dry or wet weight of the biological samples, total proteins, or DNA content. With plasma or sera, the biomarker values are also expressed in concentration (e.g., $\mu g/L$ or nmol/mL). For tissues, normalization is achieved by simply dividing the endpoint value concentration by the biomass value. When the analyte value is small compared to the biomass, the ratio is assumed constant in most cases. However, there are situations where the ratio could change and produce correction artifact. For example, in amphipods the brain neurotransmitter acetylcholine is relatively constant but muscle mass could increase drastically as the size of the organism readily increases. This could lead to decreased acetylcholinesterase activity (expressed rate of reaction/mg proteins) as the size of the organism increases. In this case, normalization induces an apparent decrease in acetylcholinesterase activity because the total protein content (from muscles) readily increases. In another case, the levels of vitellogenin (VTG)-like proteins in female gonad tissues readily increases during gametogenesis in bivalves reaching approximately > 50% of the total protein biomass in the gonad. In this situation, normalization against total protein could lead to over normalization artifacts. In these situations, normalization could be achieved by other means that will be presented in the following section: residual extraction and stepwise addition methods. These two methods could handle larger variations of analyte and biomass changes compared to the simple ratio procedure.

1.8.1 Residual Extraction Method

This methodology is based on determining the actual slope between the analyte changes and the animal's biomass measurement and within a given treatment or absence thereof (Figure 1.4). The residual value from the linear

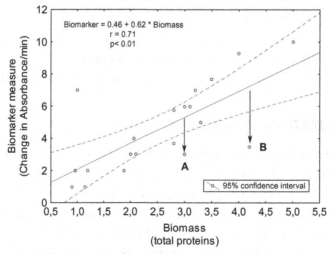

Figure 1.4 Biomass normalization using the residual extraction method.

regression between the biomarker and biomass is the difference between the measured biomarker responses and the predicted value from the linear relationships. This is exemplified by the arrows shown in Figure 1.4. This difference represents the biomarker response that was not explained (the error) by changes in biomass and constitutes the corrected biomarker value for biomass. This methodology is suitable when the linear relationship is statistically significant ($p < 0.05$) and should be applied for each treatment or exposure group ideally, but a general (all groups) trend could be produced provided the slopes are the same. This procedure could be compared to analysis of covariance where biomarker responses are compared against the various treatment groups with the biomass as the covariable. This would be another equivalent means to account for biomass variations in the biomarker response. When the slope against biomass is not significant, no normalization against the measured biomass should be made.

As an example, consider the data presented in Figure 1.4. The biomarker (enzyme activity) was determined in a given tissue homogenate extract. The y-axis represents the rate of reaction catalyzed by the enzyme expressed as change in absorbance per minute of reaction time. In the attempt to correct against biomass (normalization), the reaction rates for all the biological samples were plotted against total proteins (or other biomass endpoint). The slope is interpreted as the influence of protein changes on enzyme activity. Let's consider data points A and B (Figure 1.4); the measured values are 2.8 and 3.2, respectively. These values are the results of the initial amount of added material (biomass) and the effects of a particular treatment. To consider the effect of the treatment from the biomass effect, the residual value is obtained as follows: residual value = |measured value − value from the linear relation|. For A the residual value is $|2.8 - 5.1| = 2.3$, and for B the residual value is $|3.2 - 6.5| = 3.2$. The values are in absolute amount because we are seeking the difference or the value that was not explained by the linear regression with protein content. Some people will prefer to use the real value if it makes sense that it could take a negative value and others use the squared value. This method could be applied especially in cases where the slope between the measured biomarker and the biomass is suspected to change or differ in some way (type of tissues, condition of the organisms, etc.). We notice that some points fall within the 95% confidence interval, which means the measured value is statistically similar to a protein content effect and provides insights into the treatment effects. Indeed, the data points outside this interval could be the result of the treatment where an induction or inhibition of the activity would be significant. If all the data points fall within the 95% confidence interval, then we could assume the changes in biomarker measurement are explained only by variation in biomass. In the extreme, if no correlation is obtained then the biomass endpoint may not be suitable and cannot be used to normalize the biomarker for biomass. In this case, another biomass measure such as tissue dry or wet weight could be used. Another approach would be to use the step-wise addition method described next, which represents a more fundamental method of normalization that is based on the methodology for the biomarker evaluation itself.

1.8.2 Stepwise Addition Method

The stepwise addition method determines the rate of increase of the biomarker in respect to added sample volume. The slope between the biomarker and pro-tein biomass is replaced by the slope between the increased biochemical signal and the fold addition of volume added. The increase in biomarker signal/fold added volume ratio will gradually increase as the tissue is enriched with the biomarker and will reach the maximum (theoretical) limit equivalent to the pure biomarker. The rate of increase could also be compared with a tissue reported to contain high levels of the biomarkers if no standards are available. For example, VTG levels in gonad tissues could represent more than 50% of "soluble" proteins in some bivalves' gonad homogenates and could serve as a positive control. With this approach there is no need to divide the biomarker with a biomass metric.

For example, the evaluation of VTG by the indirect alkali-labile phosphate (ALP) assay is given. VTGs are highly phosphorylated high molecular weight proteins that could be determined on the basis of ALP contents. We spiked rainbow trout liver homogenates with VTG extract from eggs to give VTG in the liver extract representing 0, 0.2, and 1% of the total protein content. In this case, normalization of ALP with total protein concentration in the sample will lead to over normalization problems because of autocorrelation effects. The increase of ALP was determined by adding 0, 2.5, 5, 7.5, and 10 µL (0, 1-, 2-, 3-, and 4-fold) of the sample during the phosphate assay in liver and VTG-spiked liver samples. The increase in ALP per added volume ratio was also determined in the VTG-rich egg extract.

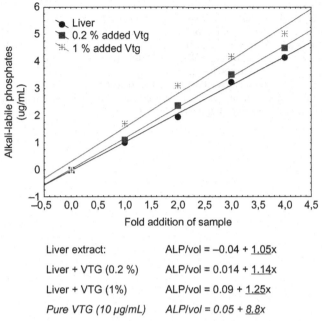

Liver extract:	ALP/vol = −0.04 + <u>1.05</u>x
Liver + VTG (0.2 %)	ALP/vol = 0.014 + <u>1.14</u>x
Liver + VTG (1%)	ALP/vol = 0.09 + <u>1.25</u>x
Pure VTG (10 μg/mL)	*ALP/vol = 0.05 + <u>8.8</u>x*

Figure 1.5 Demonstration of the stepwise addition method for normalization.

If the pure biomarker is available then the slope could be determined as shown previously and could serve to extrapolate the levels of VTG in the liver sample. The pure standard could be mixed with the sample extract to take into account any matrix effects. In the liver extract, a slope of 1.05 was obtained. Pure VTG at 10 μg/mL produced a slope of 8.8, then the level of VTG in the liver sample is $(1.05 \times 10 \, \mu g/mL)/8.8 = 1.19 \, \mu g/mL$ of liver extract (see Figure 1.5). When no pure samples are available, then a positive control with an enriched fraction or from a biomarker from another species could be used and the data compared to controls. This approach is limited only by the capacity of the analytical method to handle increasing amounts of the biomarkers in the assay volume and the availability of suitable biomarker standard, but it requires more analyses to build the calibration curve.

REFERENCES

[1] Mayer FL, Versteeg DJ, McKee MJ, Folmar LC, Graney RL, McCume DC, et al. Physiological and nonspecific biomarkers. In: Hugget RJ, Kimerle RA, Mehrle Jr PM, Berman HL, editors. Biomarkers: biochemical, physiological, and histological markers of anthropogenic stress. Boca Raton: FL: Lewis Publishers; 1992. p. 5—85.

[2] Schulte PA, Talaska G. Validity criteria for the use of biological markers of exposure to chemical agents in environmental epidemiology. Toxicology 1995;101:73—88.

[3] Soares A, Calow P. Seeking standardization in ecotoxicology. In: Soares A, Calow P, editors. Progress in standardization of aquatic toxicity tests. Boca Raton, FL: Lewis Publishers; 1993. p. 1—6.

[4] Duval D. Manuel d'assurance et de contrôle de la qualité. Rapport scientifique, Centre Saint-Laurent, Environment Canada 1995. Sections 1—14.

[5] Menditto A, Palleschi S, Minoprio A, Rossi B, Calibotti A, Chiodo F, et al. Quality assurance in biological monitoring of environmental exposure of pollutants: from reference materials to external quality assessment schemes. Microchem J 2000;67:313—31.

CHAPTER 2

Tissue Preparation and Subcellular Fractionation Techniques

François Gagné

Chapter Outline

Tissues must be processed to allow detection of the analyte with minimal interference from a complex biological matrix. Indeed, given the complexity of life processes, tissues must be handled with care to minimize as much as possible the degradation of the analyte or the endpoint under study. The extent of tissue preparation will depend on the specificity of the assay and the species. Subcellular organites are sensitive structures and special care should be given to minimize disruption of organites during freezing, thawing, and homogenization. Organites are composed of membranes most of the time and special attention to the osmolarity (salinity) of the media should be given to prevent either hypo-osmotic or hyper-osmotic shock. During cell or tissue disruption, proteases and nucleases are inadvertently liberated, which could inactivate the biomarker of interest. Enzymes and RNA are particularly sensitive to

Biochemical Ecotoxicology
DOI: http://dx.doi.org/10.1016/B978-0-12-411604-7.00002-7

protease/ribonuclease activity so special care should be provided by working at low temperatures, working at slightly basic pH, and with the presence of Ca/Mg chelators. All these aspects will be explained in this chapter.

2.1 TYPES OF HOMOGENIZATION BUFFERS

Homogenization buffers are prepared to maintain the analyte integrity and limit its biodegradation by endogenous proteases and nucleases. Homogenization of tissues for biochemical assays usually requires the disruption of the tissue structure and cells to liberate the intracellular material. The process of homogenization could be considered as a "controlled" destruction method of cells. The choice between the homogenization buffer (and method) is based on a compromise between analyte recuperation yield and integrity of the biomarker. For example, a homogenization step to liberate a cytosolic enzyme requires lysing as many cells as possible in tissues, on the one hand, and on the other hand, increased harshness of the homogenization steps in an inadequate buffer media could favor the liberation and activity of proteases, RNase, and DNase, which could readily inactivate the biochemical target under study. A Polytron type of tissue homogenization is highly efficient in disrupting hard or tough tissues such as muscles and conjunctive tissues, but the process could generate heat (and ultrasounds), which in turn could inactivate the enzyme or chemically degrade the analyte(s) of interest. The following properties of any homogenization buffer should be taken into consideration.

2.1.1 pH

The pH of the homogenization media is selected to maintain (1) the pH for the stability of the analyte, and (2) to limit the activity of proteases, RNases, and DNases. Usually slightly basic pHs (7.5−8.2) are preferred since many of the protein and nucleic acid degradation enzymes are optimal at pH 5.5−6.5 and generally have limited impact on the analyte stability. However, many enzymes optimally work at pH < 7.5 so that the buffer concentration should not be high enough to bring down the pH during the assay or care should be taken not to irreversibly inactivate the enzyme of interest. The buffer is selected to function in the pH range and within ± 1 of the pKa of the buffer to ensure good pH control. A non-exhaustive table is provided for the most commonly used buffers in biochemistry (Table 2.1). The homogenization

Table 2.1 List of Commonly used Buffers in Biochemistry Research

Common Name	Name	pKa	pH Range
Tris	Tris-(hydroxymethyl) aminomethane chloride or acetate salt	8.06 at 25°C Temperature sensitive; adjust pH at the working temperature	7.06–9.06
HEPES	4-(2-hydroxyethyl)-1-piperazineethanesulfonic acid	7.5	6.5–8.5
MOPS	3-(N-morpholino) propanesulfonic acid	7.2	6.2–8.2
Glycine	Aminoacetic acid	pKa1 2.3 pKa2 9.5	1.3–3.3 8.5–10.5
Bicarbonate (sodium)	$NaHCO_3$	pKa1 6.35 $H_2CO_3 \rightarrow HCO_3^{-1}$ pKa2 10.35 $HCO_3^{-1} \rightarrow CO_3^{-2}$	5.3–7.3 9.3–11.3
Sodium/potassium phosphate	$Na_{x3-x}H_xPO_4$ (x is between 3 and 0)	pKa1 2.3 $(H_3PO_4 \rightarrow H_2PO_4^{-1})$ pKa2 7.2 $(H_2PO_4^{-2} \rightarrow HPO_4^{-1})$ pKa3 12.1 $(HPO_4^{-2} \rightarrow PO_4^{-3})$	1.3–3.3 6.2–8.2 —*
Acetate (sodium)	CH_3COOH/CH_3COO^-	4.75	3.75–5.75
Citrate (sodium)	NaOOC-CH2-C (-COONa,-OH)-CH2-COONa	pKa1 3.13 pKa2 4.76 pKa3 6.4	2.1–4.1 3.8–5.8 5.4–7.4

*pH buffering at >12 is seldom required especially for homogenization (never required). NaOH is used instead.

temperature should always be under 4°C to limit the general chemical activity of the sample. One of the best ways to prevent temperature degradation is suggested as follows: the tissues are frozen first and the homogenization done (either Polytron or Teflon pestle tissue grinders) with the frozen tissue directly in ice-cold homogenization buffer. This procedure is actually a combination of freeze–thaw and mechanical tissue homogenization steps. In general, a Teflon pestle tissue grinder (Potter–Elvehjem homogenizer) offers the best compromise between maintenance of intracellular component (organelles) integrity and analyte yield. Each buffer has its own characteristics and should be used on a case by case basis. For example, Tris is a salt that contributes less to the osmotic pressure of the media, but the pH is sensitive to temperature changes; citrate is also a good buffer and exhibits chelating properties for divalent elements (Ca^{2+}).

Phosphate buffer could inhibit some cytochrome P450 activities (P4503A4). The buffers are generally biologically compatible when used at low to moderate concentrations (1—50 mM).

2.1.2 Ion Chelators

Calcium, magnesium, and zinc chelators are usually added in the homogenization buffers since many exonucleases and endonucleases (RNA and DNA) and proteases need these cations for optimal activity. Ethylenediamine tetraacetic acid (EDTA) is by far the most commonly used chelator for cations such as calcium, magnesium, and many other cations such as cobalt, zinc, cadmium, iron (II), and nickel. Ethylene glycol tetraacetic acid (EGTA) is more specific for calcium and less specific for magnesium. $N,N,N'N'$-tetrakis(2-pyridylmethyl) ethylenediamine (TPEN) is a good zinc chelator for zinc-dependent endopeptidase (metalloproteinase). It also an inducer of apoptosis in cells by decreasing the levels of free zinc in cells. Citrate is considered a mild calcium and iron (II) chelator, but citrate is an intermediate metabolite and could interfere if glucose metabolism is determined in the experiment.

2.1.3 Antioxidants

Some molecules are sensitive to the redox potential in cells during homogenization and the presence of antioxidants is needed in the sample buffer. For example, thiol-containing proteins could readily oxidize to form dimers and inactivate enzyme activity if present. Analysis of sensitive compounds such as thiols (glutathione, cysteine-rich proteins, sodium nitrite, ascorbic acid, and retinoic acid) requires tissues to be prepared under reductive conditions. β-mercaptoethanol or dithiothreitol are the most commonly used antioxidants in homogenization buffers and used at concentrations between 0.1 and 10 mM. Other antioxidants such as ascorbic acid, cysteine, and glutathione could also be used, although care should be taken to not interfere in the assays because of their biogenic origins. Tris (2-carboxyethyl) phosphine is a very strong reducer with some alkylating potential and is becoming a more common reducing agent especially in the area of gel electrophoresis analysis of proteins (proteomics).

2.1.4 Proteases Inhibitors

Laboratory workers should always be aware of the presence of proteases in biological samples. Usually they are active at pH < 7 (slight acidic condition) and many of them require cations (Ca^{2+} or zinc) to work. The most fundamental way to block proteases is to maintain the tissue extracts at $0-4°C$ in slightly basic homogenization buffers (at pH 8–8.5) in the presence of chelators (EDTA or EGTA). EDTA is not able to inhibit trypsin activity, however. Phenylmethanesulfonyl fluoride (PMSF) is a general protease inhibitor that inhibits serine proteases such as trypsin and chymotrypsin. It also inhibits cysteine proteases and mammalian acetylcholinesterase. It is usually used at concentrations between 0.1 and 1 mM in homogenization buffers. Aprotinin and leupeptin are also used as protein-based protease inhibitors for trypsin and serine/cysteine proteases. However, we cannot assume their complete effectiveness for all species and precautions should always be used with new biological samples, especially when working with new species. As a general guide, working at a slightly basic pH at cold temperatures and with a combination of EDTA + another inhibitor (PMSF) usually gives good results. However, if tissues from the digestion tract are to be examined then a more complete array of protease inhibitors should be considered. Commercial protease inhibitor cocktails are available. In some case, proteases could be quenched by adding an excess amount of protein such as albumin, for example, during the isolation of cells of organites, provided that excess proteins do not interfere in the assays afterward.

2.1.5 DNase/RNase Inhibitors

The general criteria for proteases such as slightly basic pH, low temperatures, and the presence of a cation chelator (EDTA) can also be applied to nucleases in general. Diethylpyrocarbonate (DEPC) is a well-known ribonuclease or RNase inhibitor. It is used to treat water at a concentration of 0.05–0.1%, which can be removed by heating at $80-90°C$ for 15 min if needed. If DNA or RNA is to be extracted then heat denaturation at $75-80°C$ for 10 min following homogenization in ice-cold basic buffer containing EDTA (e.g., 10 mM Tris-acetate, pH 8.2, with 1–10 mM EDTA) constitutes a good method to inactivate RNases or DNases. Many commercial sources of recombinant protein-based RNase and DNase inhibitor cocktails are currently available (Qiagen, Life Technologies, Invitrogen). Chemical inhibitors of DNases such as β-mercaptoethanol (10–100 mM), EGTA and EDTA (10 mM), sodium dodecylsulfate (0.5–1%), and iodoacetate (10 mM) are also commercially available. The efficacy of each of them will depend on the nature of the tissues and species under investigation.

2.1.6 Osmolarity

Osmolarity is defined as the number of ionic species in molarity that has a characteristic range depending on the species examined. It is calculated as the sum of molar ionic species in a media, for example, 150 mM NaCl has an osmolarity of 150 mM Na^+ + 150 mM Cl^- = 300 mOsmol; 50 mM $CaCl_2$ and 5 mM $NaHCO_3$ have an osmolarity of 50 mM Ca^{2+} + 2 × 50 mM Cl^- + 5 mM Na^+ + 5 mM HCO_3^- = 160 mOsmol. In vertebrates, the plasma osmolarity values are range from 275 to 325 mOsmol. The maintenance of cells, membrane vesicles, or intracellular organelles (microsomes, mitochondria, and nuclei) is dependent on the transmembrane equilibrium of osmotic pressure to prevent hypo-osmotic or hyper-osmotic stress. For example, cell culture media are usually adjusted to 290−320 mOsmol to prevent any osmotic stress to cells. If the osmolarity is not known then a value could be determined by measuring the conductivity of cells or tissues homogenized in water if the plasma (or hemolymph) is not available. In freshwater mussels, the hemolymph has an osmolarity of 70−100 mOsmol (35−50 mM NaCl). The osmolarity of homogenization buffers or cell culture media must be suitable for the species under investigation, especially if the maintenance of membrane vesicles such as nuclei, mitochondria, lysosomes, and microsomes is required. Osmolarity is usually adjusted with NaCl, KCl, Na/KH_2PO_4, or sucrose. Compounds that do not pass the membranes (Percoll or dextrans) or readily pass the membranes (dimethylsulfoxide; DMSO) are not thought to contribute significantly to osmolarity. For example, HEPES and Tris-based buffer and large proteins (albumin) were reported to weakly contribute to the osmotic pressure.

2.1.7 Cryoprotective Agents

In some instances, a number of enzyme complexes or protein-bound membrane receptors are sensitive to freezing temperatures since crystallization occurs during freezing. The addition of cryoprotective agents that prevent or reduce crystal formation during deep freezing are required. They could be directly added in the homogenization buffer or included just before freezing to a given homogenate fraction. DMSO and glycerol are the most commonly used cryoprotective agents. The concentration of DMSO is usually ∼ 10−30% and glycerol between 10 and 50%. Other cryoprotective agents such as sucrose, glucose, ethylene glycol, and propylene glycol could also be used.

2.2 HOMOGENIZATION METHODS

A selection of homogenization methods is proposed in this section. The efficacy of homogenization methods is a balance between yield and sample integrity as mentioned earlier. On the one hand, mild homogenization methods are generally well suited for delicate structures (membrane vesicles or large DNA macromolecules) but the yield could be low. On the other hand, harsh homogenization methods could significantly increase the yield but at the expense of sample integrity. For example, a Polytron homogenizer could process sturdy tissues but at the expense of destroying cellular organelles (e.g., mitochondria or genomic DNA). In some cases, a combination of homogenization procedures could be used to improve yield with minimal loss of biological integrity.

2.2.1 Teflon Tissue Pestle

The Teflon pestle tissue grinder (Potter–Elvehjem) is by far the best method to isolate cellular organelles. The Teflon pestle diameter tightly fits to the tube so that a tight space forces the tissues to pass through this space during the up and down passes (Figure 2.1). The Potter pestle also turns by a motor (a piercing drill is sometime used) at varying speeds. Usually 5 to 10 strokes or passes are required to homogenize a sample.

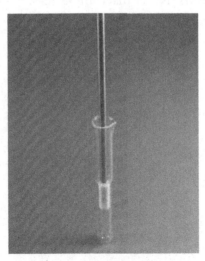

Figure 2.1 A Teflon pestle tissue grinder.

This method is convenient for relatively small tissues (about 10 mg–10 g) sizes. Larger sizes are required to mince tissues into smaller pieces and homogenize them in sequence. Homogenization should be done at <4°C, i.e., in ice-cold buffer, with slightly frozen tissues, and the Teflon pestle tube surrounded by ice or placed in a 4°C chamber.

2.2.2 Polytron

The Polytron tissue grinder (Figure 2.2) represents a stronger homogenization technique and is used for robust tissues such as cartilage and muscle. It is considered a harsh process that can lead to important subcellular and macromolecular damage. Rotary blades, which are tightly fitted in a metallic tube, form the basis of the tissue-grinding process. This process, however, generates heat and ultrasounds so care should be taken to keep the tissue frozen during homogenization and to limit macromolecule and organelle breakdown. Polytron tissue grinders are sold as hand-held instruments, which makes them easy to handle.

2.2.3 Ultrasounds

Tissues and cells could be readily disrupted in ultrasonic water baths. More adapted to disrupt cellular membranes and macromolecules in cells, this method could generate significant amounts of heat in the process so it can denature proteins and DNA/RNA macromolecules.

Figure 2.2 Polytron-based tissue grinder.

2.2.4 Freeze—Thaw

The freeze—thaw process represents a "mild" homogenization procedure and it is usually used in combination with other homogenization methods. It consists of quickly freezing at $-85°C$ and thawing at $4°C$ in sequence (the tubes could be thawed quickly at $6-10°C$ in a water bath for 10 min and kept at $4°C$). Usually two to three freeze—thaw cycles are necessary. Most of the time, tissues are freeze—thawed once and homogenized using a Teflon pestle tissue grinder (two to four passes) for best results. However, the activity of some enzymes are sensitive to repeated freeze-thaw cycles. Drops in activities in the order of 30% could be observed with some enzymes. Thus, enzyme activities in samples should be compared with the same number of freeze-thaw cycles (usually <3 cycles).

2.2.5 Subcellular Fractionation by Differential Centrifugation

Separation of the intracellular component is usually achieved by differential centrifugation. A general procedure is shown in Figure 2.3 that can be adapted

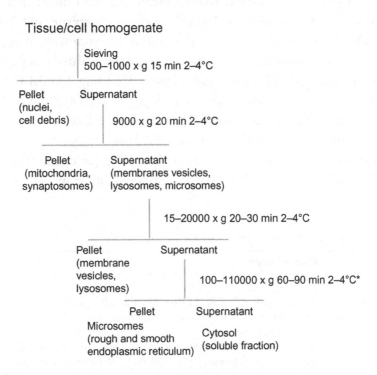

* Require an ultracentrifuge

Figure 2.3 General centrifugation scheme for the isolation of intracellular fractions.

depending on the tissue type or test species under investigation. The homogenate usually contains tissue fibers, cell debris, and intact cells, which are removed by a sieving (50–500 μm) procedure such as metallic or plastid grids or "cotton-cheese" wool cloth. This step is optional if the isolation of pure nuclei is not an objective. The homogenate is centrifuged between 1000 and $1500 \times g$ for 15 min to isolate large cell debris and nuclei. The supernatant is then centrifuged at $9000 \times g$ for 20–30 min at 2–4°C to obtain the crude mitochondrial pellet and the supernatant. With nerve tissue homogenates, the pellet also contains crude synaptosomes from nerve cells (from nerve terminals). Mitochondria and synaptosomes could be separated further by sucrose density gradient centrifugation as explained in Chapter 9. The supernatant could be centrifuged between 15 and 20,000 $\times g$ for 20 min to isolate membrane vesicles and lysosomes (pellet). Finally, the supernatant is ultracentrifuged at 105,000 $\times g$ for 60–90 min at 2–4°C to pellet the microsomes (smooth and rough endoplasmic reticulum) and the cytosol (supernatant). The microsomes are usually resuspended in ice-cold 10 mM HEPES-NaOH buffer, pH 7.4, containing 150 mM NaCl or KCl (or 100 mM sodium phosphate, pH 7.4), 1 mM EDTA, 1 mM dithiothreitol, and 20% glycerol as a cryoprotective agent. If no ultracentrifuge is available then the microsomes could be isolated using the calcium precipitation method [1]. This procedure consists of adding 10 mM $CaCl_2$ to the 9000 or 12,000 $\times g$ supernatant and centrifuge at 15,000 $\times g$ for 30 min at 2–4°C. The aggregated microsomes from calcium are found in the pellet and resuspended in the same type of buffer as described previously.

Marker enzymes could be used to assess the purity (or contamination by) of the subcellular fraction (Table 2.2). The enzymes are either specific or expressed

Table 2.2 Marker Enzymes for the Major Organelles

Organelle	Marker Enzymes
Nuclei	
Plasma membrane	5′-nucleotidase Na/K-ATPase Alkaline phosphatase
Mitochondria	Cytochrome c oxidase Succinate dehydrogenase
Lysosomes	Acid phosphatase
Microsomes	NADPH–cytochrome P450 reductase, cytochrome P450-related enzymes (benzo[a]pyrene hydroxylase, dibenzoylflourescein dealkylase, ethylmorphine-N-demethylase)
Cytosol	Glutathione S-transferase

at high activity in the target organelles. This represents a quantitative means to assess the isolation and enrichment procedure for subcellular fractions [2].

2.2.6 Conservation of Biological Samples for Biomarkers

Biological samples are usually conserved frozen for long-term storage. For long-term storage ($>$1 month), samples are usually frozen at $-85°C$. Some laboratories use dry ice for freezing but the temperature is at $-78.5°C$. For short-term freezing ($<$1 month), a freezing temperature of $-20°C$ is acceptable. Especially for enzymes, activity is lost after each freeze—thawing cycle. In general, a 20—30% loss in enzyme activity occurs after each freeze—thaw step, especially in the absence of cryoprotective agent like glycerol, but this should be checked with each enzyme. Membrane-bound enzymes and quaternary enzyme assemblages are particularly sensitive to repeated freeze—thaw cycles. Try to organize your work so only a maximum of one or two freeze—thaw cycles is required.

REFERENCES

[1] Hamilton RL, Moorehouse A, Lear SR, Wong JS, Erickson SK. A rapid calcium precipitation method of recovering large amounts of highly pure hepatocyte rough endoplasmic reticulum. J Lipid Res 1999;40:1140—7.
[2] Padh H. Organelle isolation and marker enzyme assay. In: Goldman CA, editor. Tested studies for laboratory teaching, vol. 13. Proceedings of the 13th Workshop/Conference of the Association for Biology Laboratory Education (ABLE), 1992. p. 129—46.

CHAPTER 3

Preparation and Maintenance of Live Tissues and Primary Cultures for Toxicity Studies

Brian Quinn
University of West Scotland, Scotland, UK

Chapter Outline

3.1 GENERAL INTRODUCTION

In vitro toxicology studies are conducted using parts of an organism (tissues or cells) that have been isolated and grown or maintained under controlled conditions. This technique is also known as tissue or cell culture. As living organisms are extremely complex functional systems, the primary advantage of this technique is that it allows the investigation of individual components of this complex system and the study of fundamental biological functions in cells. However, this simplification of these very complex systems is also a significant disadvantage as often the results from *in vitro* experimental studies are not necessarily predictive of effects in whole organisms, and can sometimes lead to

Biochemical Ecotoxicology
DOI: http://dx.doi.org/10.1016/B978-0-12-411604-7.00003-9

misinterpretation of results at the *in vivo* level. In other words, *in vitro* test results should be validated with the whole animal in relation to the objective of the investigation under way. However, once the constraints of the *in vitro* system (primarily revolving around the loss of cellular-specific function in respect to the tissue of collection) are realized these techniques could be powerful screening toxicity tests to determine the mode of action of xenobiotics at the cellular (fundamental) level. Indeed, primary cultures of fish hepatocytes were recognized as model systems in toxicology research for many years since they retain the metabolic capacity of the liver from many days up to 1 week depending on the culturing conditions.

In vitro techniques are particularly valuable in toxicity testing, where they are commonly used for a number of ethical (reduction in the number of test organisms), scientific (cells provide a key level of biological organization), and economic (can provide rapid, low cost, and reliable screening tools) reasons [1]. Indeed, *in vitro* systems are recognized alternatives to *reduce* the use of animals in toxicity testing, *refine* toxicity evaluation, and *replace in vivo* studies (the so-called three Rs of alternative tests). Cell and tissue culture approaches represent a powerful means for the investigation of the mechanism of toxicity on individual cells removed from the processes of the entire organism [2]. The use of immortalized cell lines is very common in toxicity testing when studying basic cellular functions, since these cells tend to be undifferentiated (i.e., loss of tissue-specific characteristics) over time. The fields of fish primary cultures and cell lines are much more developed than those for aquatic invertebrates. For this purpose, the described method is focused on the preparation of primary cultures from aquatic invertebrates, namely mussels. In vertebrates (fish, birds, and mammals), primary cultures of cells are mostly isolated based on the double perfusion methodology developed by Klaunig et al. [3], which was based upon mammalian techniques [4]. At the first step, tissues are perfused with a calcium chelator to loosen the intercellular matrix since calcium is involved in cell adhesion and cohesion. The second step consists of perfusing the tissues with proteases (e.g., collagenase, trypsin, or Pronase) to liberate further the cells from the intercellular matrix. To date limited success has been achieved with bivalve primary culture preparations where bivalve cell lines have yet been established, and only one molluscan cell line developed, the Bge cell line [5]. For this reason primary cultures of cells in suspension or of tissue explants have been more commonly used for bivalve toxicity testing *in vitro*. Notwithstanding, isolated cells from various bivalve species have been used in many toxicological studies, particularly from the digestive gland [6–8], hemolymph [9–14], and gills [8,15,16]. These tissues are important in toxicity testing as they are involved in the detoxification and defense mechanisms and are in constant contact with the environment (respectively). Other tissues investigated include the mantle [17–19] and heart cells [20].

The ultimate aim in tissue culture is the ability to reproduce the *in vivo* mechanisms of metabolism and detoxification/transformation as close as possible for toxicity testing. However, this has not yet been possible, especially for cell lines, as removing tissues and cells from the complexities of the whole organism (and tissue organization) changes the ability to observe toxicity effects *in vitro*. Currently the area of invertebrate tissue culture is still in its infancy with many of the fundamental questions on the nutritional requirements and metabolism of these animals at a cellular level not completely answered. Only time and experience will contribute to better these *in vitro* testing systems. In many studies, isolated cells (particularly hemocytes) exposed for short periods (up to 24 h) were often used and are considered as primary cultures. This chapter focuses on a previously published technique [2] and describes in detail the establishment of primary cultures from the freshwater zebra mussel (*Dreissena polymorpha*) and their use in acute cytotoxicity testing (cell suspensions) and chronic biomarker expression (explants). The usefulness of bivalve molluscs as bioindicators has long been established due to their high filtration rate; ability to accumulate and bioconcentrate toxicants; widespread distribution and abundance; and their primarily stationary, benthic life cycle. This chapter outlines all the major steps necessary to develop a technique for establishing primary cultures and can relatively easily be adjusted for use with other bivalve species. The descriptions given below for the composition of the various solutions including the media, sterile buffer solution, and antibiotic and Pronase solutions can all be relatively easily modified to better suit other (e.g., marine) bivalve species by the adjustment of the osmolarity and chemical composition to closely resemble the hemolymph of the animal tested.

3.2 REAGENTS REQUIRED AND SOLUTION PREPARATION

Sterile cell culture water, phenol red sodium salt, HEPES sodium salt (minimum 99.5% titration), kanamycin sulfate (from *Streptomyces kanamyceticus* 767 µg mg^{-1} dry base), sodium chloride (NaCl), gentamicin (50 mg mL^{-1}), Pronase (from *S. griseus*), 3-(4,5-dimethyl-2-thiazolyl)-2,5-diphenyl-2H-tetrazolium bromide (MTT), trypan blue solution (0.4%), isopropanol (minimum 99%), and dimethylsulfoxide (DMSO; CAS number 67-68-5; purity $\geq$ 99.9%) were obtained from Sigma-Aldrich (Steinheim, Germany). Fetal calf serum (FCS), L-glutamine (200 mM), penicillin-streptomycin (Pen-Strep; 5000 IU mL^{-1}–5000 µg mL^{-1}), and Leibovitz L-15 Medium were purchased from Gibco. Hydrochloric acid solution (HCl 6 N) was purchased from Fluka, while ethanol came from Merck (Darmstadt, Germany). For each solution the pH and osmolarity were checked and adjusted to resemble that of the animal's hemolymph, in this case for the zebra mussel, which was found to be 7.5 and 80–100 mOsm, respectively.

Culture media: The culture media in particular is developed to resemble as closely as possible the hemolymph of the animal cultured. In the current example of the zebra mussel, media was developed using 15% Leibovitz L-15 media consisting of (1L) the following: 150 mL Leibovitz L-15, 5 mL Pen-Strep (5000 IU mL^{-1}–5000 µg mL^{-1}), 2 mL gentamicin (50 mg mL^{-1}), 0.01 g kanamycin (759 µg mL^{-1}), 0.01 g phenol red, 843 mL sterile water, and 2.38 g HEPES. The osmolarity and pH were regulated to 80–100 and 7.5 mOsm, respectively, and the media sterile filtered (0.22 µm; Millipore) and stored for up to 6 months at $-20°C$. Once defrosted to room temperature 100 mL FCS (10% final volume; Gibco) and 10 mL L-glutamine (200 mM; Gibco) were added and the osmolarity and pH checked and adjusted accordingly.

Sterile buffer solution (SBS): To 1 L sterile cell culture water or deionized water dissolve 2.32 g NaCl (Sigma) and the osmolarity (80–100 mOsm) and pH (7.5) checked and adjusted where necessary. Deionized water can be sterilized by filtering through a 0.2 µm pore filter under sterile conditions (sterile laminar flow hood) or by heating at 80–90°C for 30–45 min.

Antibiotic solution (4X): The antibiotic solution (1 L) contains the following: 972 mL of sterile water (Sigma), 20 mL Pen-Strep (5000 IU mL^{-1}–5 000 µg mL^{-1}), 80 µg mL^{-1} gentamicin (50 mg mL^{-1}), and 40 µg mL^{-1} kanamycin (759 µg mL^{-1}). This antibiotic solution ($\times$ 4) was subsequently diluted one-quarter with SBS.

Pronase: A Pronase solution (from *S. griseus*) was made at a concentration of 0.025% using 75 mL SBS, 25 mL antibiotic solution ($\times$ 4), and 25 mg Pronase (4 IU mL^{-1}).

Trypan blue solution: Prepare a 0.4% solution by suspending 0.2 g of trypan blue in 50 mL of SBS. Mix thoroughly and filter the solution on a

Whatman #1 paper to remove aggregates. This solution could be also purchased commercially (e.g., Sigma–Aldrich).

MTT solution (for cell viability): A stock solution was prepared by dissolving MTT into SBS at a 5 mg mL^{-1} final concentration. This solution was sterile filtered through a 0.22 μm filter and stored at 4°C.

Acidic isopropanol: Dilute concentrated HCl in isopropanol to have a final concentration of 40 mM; for example, 33 μl of concentration HCl (12 N) in 100 mL isopropanol. Wear protective gloves, clothing, and eyewear and work in a fume hood.

Culture vessels: The choice of culture vessel will depend on what the samples are to be used for (see Section 4). The cells in suspension for use in toxicity testing are usually seeded in 96-multiwell culture plates. Cell adhesion can be encouraged and enhanced by the use of tissue culture-specific plates with the culture surface treated or coated with an appropriate solution (e.g., collagen) to enhance cell adhesion (e.g., BD Falcon BD Primaria, Corning Costar, or Sarstedt Cell + 96-well cell culture plates) or the use of a specific cell adhesive (e.g., BD Biosciences Cell-Tak cell and tissue adhesive). Tissue explants for use in chronic biomarker expression were placed in pre-autoclaved 1.5 mL Eppendorf tubes. The use of these tubes allowed for the tissue homogenization using a pellet pestle motor (Kontes) with no sample loss.

3.3 PROCEDURE

3.3.1 Animal Collection and Maintenance

As bivalves are filter feeders and can contain large numbers of contaminants they should be depurated in the lab before culturing. Animals should be as healthy as possible for tissue culture, with animals taken from the wild replaced every 2 weeks with new fresh individuals. Animals should be collected with minimal interference by cutting the byssal threads (not tearing them), quickly transported back to the lab, and maintained in suitable holding facilities that reflect their natural environment (e.g., temperature, photoperiod, salinity, etc.). Thoroughly clean the mussels and holding facilities to help reduce contamination. Aquaria should be cleaned regularly (three times per week), the water replaced, and animals fed an appropriate diet consisting of an algal/phytoplankton cocktail commercially available in most pet shop stores.

3.3.2 Preparation of Primary Cell Cultures

All cell culture procedures should be carried out under sterile conditions in a laminar flow hood (Class II microbiological safety cabinet) exclusively used for cell-culture purposes. All glassware and equipment used for culturing is autoclaved before use and all solutions sterile filtered through a 0.22 μm filter (Millipore). Good aseptic technique should be maintained throughout to ensure no contamination. The entire procedure for the culture of cells in suspension and tissue explants from the gill and digestive gland is summarized in Table 3.1 and described in detail in the following sections.

3.3.2.1 Dissection and Decontamination

The bivalve shell of each individual is scrubbed clean under running water and mussels placed in sterile tap water. Once transferred into the laminar flow the mussels are placed in antibiotic solution ($\times 2$) for 2 h, allowing them to filter the antibiotic solution through their gills and aid decontamination. Attempts should be made to keep the mussels at their preferred temperature (partially on ice). Before dissection, the bivalve shells are rinsed in ethanol 70% and allowed to dry. Using a scalpel the bivalve shells are opened and the visceral mass rinsed with 10 mL of sterile water. Following this, the tissues are dissected out and kept in separate Petri dishes containing SBS. Tissues from several animals can be pooled together, but different tissues should be kept apart. Trim tissues to get as pure a tissue sample as possible. Cut tissues up into $1-2$ mm^2 pieces and add to a Petri dish containing $\times 4$ antibiotic solution for 30 min. Using sterile forceps remove each piece of tissue and replace in $2\times$ antibiotic solution (diluted in SBS) for 20 min and later in $1\times$ antibiotic solution for 10 min. This process can be speeded up by using 15 mL tubes on a roller plate and replacing the antibiotic solution using a pipette following gentle centrifugation (500 rpm for 2 min). Finally, rinse tissue pieces in SBS under sterile conditions.

3.3.2.2 Tissue Dissociation

There are several methods available for tissue dissociation including mechanical (cutting, mincing, and sieving), chemical (chelating agent, e.g., EDTA), and enzymatic (e.g., collagenase, trypsin, and Pronase). Enzymatic dissociation is generally seen to be the best approach leading to higher cell viability and yield. For zebra mussel, Pronase (0.025% from *S. griseus*) proved to be the most successful dissociation method [2]. Following antibiotic treatment the different

Table 3.1 A Summary of the *in vitro* Technique Developed for the Culture of Both Cell Suspensions and Explants of Tissues from the zebra mussel (*D. polymorpha*)

Dissection and decontamination

- Scrape mussel clean under running water, place in sterile water
- Place in antibiotic solution $(2 \times)$ for 2 h in laminar flow partially on ice
- Rinse mussel in ethanol 70°C, allow to dry
- In laminar flow, rinse mussel with 10 mL sterile H_2O, dissect tissues
- Place tissues (digestive gland, gill, gonad) in SBS
- Trim tissue to ensure as pure a sample as possible
- Rinse in Petri dish of SBS, cut tissue into 1–2 mm^2 pieces
- Antibiotic solution (separately)

$\times 4$ (10 mL of anti-b sol.)	30 min
$\times 2$ (5 mL of antibiotic solution + 5 mL SBS)	20 min
$\times 1$ (2.5 mL antibiotic solution + 7.5 mL SBS)	10 min

- Rinse in SBS

Dissociation with Pronase

- Add tissue to 0.025% Pronase in SBS (12.5 mg in 50mL) with antibiotic ($\times 1$)
- Keep separate organs from different animals in separate tubes
- Store at slight angle at 4°C for specified time
- Filter sample liquid through autoclaved gauze 60 μm (slowly) into centrifuge tube
- Rinse tube and gauze with buffer solution
- Centrifuge filtered liquid for 3 min at 1200 rpm
- Remove liquid. Add SBS and recentrifuge, 3 min at 1200 rpm ($\times 2$)
- Remove supernatant
- Add media, mix cells using pipette, calculate cell density using a hemocytometer, adjust media volume if needed
- Place cell suspensions in culture, 1 mL in Petri (8.8 cm^2) and 0.3 mL in 24-well multiwell plate, add 0.5 and 0.2 mL media, respectively, after 24 h 15°C incubation

Explant

- Take tissue material from gauze and place in Petri (8.8 cm^2) with buffer solution
- Cut up tissue into 1 mm^2 using scalpel
- Place 10–12 explants in Petri dish; if drying place drop of media on explant
- Leave for 10 min
- Add 1 mL of media and slowly immerse explant, add 0.5 mL after 24 h incubation
- Incubate at 15°C

This technique is adapted from [2].

tissues were placed separately into 15 mL sterile tubes containing 10 mL Pronase and left to dissociate on a rocker at 4°C for a specified time (12, 16, and 40 h for the digestive gland, gills, and mantle, respectively).

3.3.2.3 Cell Suspension Culture

After dissociation, tissue samples in Pronase solution are slowly filtered through a funnel containing autoclaved gauze (60 μm) into sterile centrifuge tubes to separate out the cells in suspension that flows through the gauze from the tissue explants that are trapped in the gauze. The cells are centrifuged for 3 min at 1200 rpm at 15°C. The supernatant is removed, the cells resuspended in SBS, and the sample recentrifuged. This process is repeated twice more to wash the Pronase solution from the cells. The cells were then resuspended in the modified Leibovitz L-15 medium with cell density and viability calculated using a Neubauer hemocytometer. Generally a cell density of $\sim 1 \times 10^6$ cells mL^{-1} is achieved, but this may vary depending on the endpoint to be investigated and the initial amount of tissue dissociated. Cell viability, using the trypan blue exclusion assay (explained below), is normally >90% and must be >80% at the very least; briefly 50 μl of cell suspension is mixed with 50 μl of trypan blue solution, wait for 10 min and examine in a hemocytometer under 200–400 × microscope. Cells are seeded in the appropriate culture vessel and incubated at 15°C in a thermostatically controlled incubator. This procedure has allowed gill and digestive gland cells from the zebra mussel to remain viable for at least 15 days in culture at 15°C without any medium change [2].

3.3.2.4 Tissue Explant Culture

Tissue explant pieces were picked out from the gauze using forceps and placed into a Petri dish containing SBS, keeping the different tissues separate. Larger explant pieces were cut up into smaller sizes (~ 1 mm^2). Gently shake the Petri dish to wash the Pronase from the explants. The explant pieces were picked out using forceps and placed into a Petri dish of SBS on ice for a second wash. The explants are then immersed in media in culture vessel appropriate for their intended use (see Section 4.2) and incubated at 15°C in a thermostatically controlled incubator. These explants can remain viable for up to 14 days in culture. Explants that have not been subjected to dissociation can also be used. In this method the tissues are dissected from the animal and decontaminated as above. Once this step is complete any larger explant pieces are further cut into 1 mm^2 pieces, immersed directly in media in the appropriate culture vessel for their intended use, and incubated as above.

3.4 APPLICATION TO TOXICITY TESTING

3.4.1 Cytotoxicity Evaluation

Cells in suspension are conveniently used for cytotoxicity testing to measure the lethal effects of a chemical or pollutant. Often cells from the same tissue (e.g., gill or digestive gland) from several different individuals are pooled in order to provide enough sample to cover the need for replication in a toxicity test. Cell densities may also vary due to the level of replication needed. *In vitro* toxicity assessment should be carried out following the same principles as *in vivo* exposures, with the appropriate use of controls, solvent controls, and replication in the experimental design. The digestive gland and gills are the tissues most commonly used in toxicity testing due to their involvement in detoxification and continuous exposure to the environment, respectively. Hemocyte cells have previously been used in short-term cytotoxicity tests to investigate the effects of pollutants [8] and pathogens [21]. This offers a more straightforward protocol as no tissue dissociation is required. Hemolymph just needs to be extracted from the animal, with the hemocytes separated from the hemolymph by centrifugation (e.g., 200 g for 10 min at 15°C) and resuspended in media at the appropriate cell density [8,22]. It is important to note that in many cases bivalve cells in culture will not necessarily adhere to the surface of the culture vessel. This is important as a change of media during the exposure of these cells to a chemical in a toxicity test can result in uneven cell numbers leading to errors in the dose—response relationship. For this reason it is not recommended to change the media for cell suspensions without a centrifugation step (e.g., at $500 \times g$ for 5 min), unless you are confident that your cells are sufficiently adhered to the culture vessel. Normally for efficiency a 96-well plate is used as the culture vessel and these plates could be centrifuged with adapters. This format allows for the testing of numerous chemical concentrations with good replication. Once cell viability and density have been checked the appropriate amount of chemical to be tested can be added to the cells in suspension, incubated (at 15°C) for the appropriate time (usually 24, 48 or 96 hours), and the cytotoxic effects measured. The most common methods for assessing cytotoxicity are the MTT and the trypan blue exclusion tests as detailed below and summarized in Figure 3.1.

3.4.1.1 Cell Viability Assessment

The MTT cell viability assay is a colorimetric assay based on the method developed by Mosmann [23] for lymphocyte viability assessments, which determines the mitochondria activity, hence providing information on cellular energy metabolism. This technique was subsequently adapted to measure the viability of bivalve [8,20] and other invertebrate [24] cells in culture. In the following example cells are suspended in 100 μL (total volume) of media and incubated in a 96-multiwell plate. The protocol is as follows: 10 μL of MTT stock solution was added directly to 100 μL of cells in suspension into the 96-well plate. The cells were incubated for 4 h at 15°C in the dark. The reduction of MTT by electrons produced in mitochondria leads to the formation of an insoluble formazan product, which is solubilized by the addition of 100 μL of acidic isopropanol. After shaking the plate for 10 min, the absorbance of the solution is taken at 560 nm using a spectrophotometer with a reference wavelength of 620 nm to measure background absorbance, subtracted from the 560 nm reading.

The trypan blue exclusion test assesses cell membrane integrity providing information on cell viability [25] and has been performed by many authors both on hemocytes in culture [26] and other invertebrate cell types [2,8]. This technique offers a very simple method of using a Neubauer hemocytometer to assess cell viability based on the uptake of the blue dye by the damaged (dead) cells. There are numerous techniques published online describing how to count cells using a hemocytometer including a method published by Strober [27]. Despite being a simple technique, it is relatively time-consuming and can be somewhat subjective, depending on the eye of the examiner. In a 96-well plate, 25 μL of trypan blue solution is added to 25 μL of cell suspension. Gently mix the solution by slow aspirations with a pipette. Allow the solution to stand for at least 5 min, but no longer than 15 min as prolonged staining might lead to unspecific staining. A 10 μL volume is deposited into the Neubauer hemocytometer (with the coverslip in place). Carefully fill both chambers of the hemocytometer by capillary action. Using an inverted microscope the number of dead (blue) and live cells are counted in the 1 mm center square and four

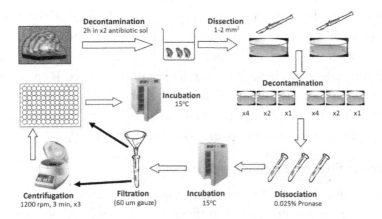

In vitro cell suspension ACUTE ENDPOINTS

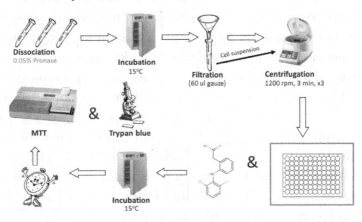

In vitro Chronic Toxicity Exposure

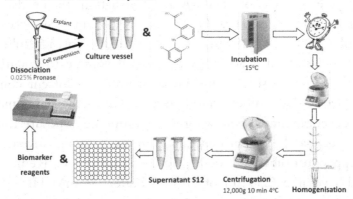

Figure 3.1 A visualization of the method developed for the culture of cell suspensions and tissue explants from the zebra mussel (*D. polymorpha*) and their use in acute and chronic toxicity testing, respectively.

1 mm corner squares. The cells touching the middle line of the top and left perimeter of each square are included in the count (but not those touching the middle line of the bottom and right sides). Try to keep a separate count of viable and nonviable cells using two handheld tally counters. Repeat the procedure for chamber 2; if the cell count is <200 and >500 cells in the 10 squares, repeat the procedure adjusting the cell density by the appropriate dilution factor. Repeat the counting procedure to ensure accuracy.

3.4.2 Chronic Effects (Biomarker Expression)

Chronic endpoints investigating the expression of proteins or enzyme biomarkers may also be investigated *in vitro* as these *in vitro* toxicological studies can provide a good mechanism to study metabolic pathways, which are difficult to investigate *in vivo*. Several reported studies use homogenized digestive gland or hemocytes in short-term (from up to 24 h) cultures to investigate the expression of different esterases, lysosomal membrane stability (indicator for general toxic stress, measuring the integrity of cell membranes), and immunological responses [6,12,28,16]. However, it is also useful to investigate the potential toxicological effects in tissue explants as these tissue fragments may better preserve the tissue-specific metabolism associated with the original tissue and may be more representative of the *in vivo* whole animal. Tissues that can be maintained *in vitro* for longer periods may also provide better models for chronic toxicity exposure as chronic effects may need a longer exposure time to be induced. Below (and summarized in Figure 3.1) is a detailed method for the culture of tissue explants from the zebra mussel for use in investigating biomarker expression.

Following the disinfection and tissue dissociation (if needed) protocol described in Section 3: explants ($n = 3$) are placed into autoclaved 1.5 mL Eppendorf tubes containing 190 μL of culture media. To each tube, 10 μL of the chemical to be tested is added giving a total volume to 200 μL (dilution factor of 20). Always be sure to include a control and solvent control when appropriate. Gently shake tubes using a vortex. This is the start time of exposure to the test compounds. Following the appropriate exposure time (usually 24 h or 96 h but may be longer depending on the endpoint investigated) centrifuge the Eppendorf tubes at $2000 \times g$ for 3–5 min at 4°C. Remove supernatant and add a volume (250 μL) of HEPES buffer (10 mM HEPES, pH 7.4, containing 100 mM NaCl, 0.1 mM EDTA, and 0.1 mM dithiothreitol). Keep sample cool (on ice) at all times. Homogenize the tissue in the Eppendorf tube using a pellet pestle motor (Kontes). Ensure total explant homogenization. Centrifuge sample 12,000 × g for 10 min at 4°C. Collect and transfer 160 µL of supernatant (S12) into a 500 µL Eppendorf tube for later biomarker analysis and 40 µL S12 into a 500 µL vial for S12 later total protein analysis. Freeze at −80°C until biochemical analysis of the biomarker (for example, determination of glutathione S-transferase activity as explained in Chapter 7) is complete.

3.5 CALCULATIONS/DATA HANDLING

3.5.1 Calculating Cell Density and Viability Using the Neubauer Hemocytometer

Cell density: Following the technique described in Section 4.1, let's suppose 200 cells were counted in the five small squares of the hemocytometer. Each square has a total volume of 0.004 mm^3 (area of 0.04 mm^2 and depth of 0.1 mm). For five squares the combined volume is 0.02 mm^3 (0.004 mm^3 × 5). Therefore there are 200 cells in a volume of 0.02 mm^3, which gives 200/0.02 = 10,000 cells per mm^3. As there are 1000 mm^3 in a cubic centimeter (same as milliliter), the cell count is 10,000,000 cells/mL. As the volume of media added to the cells is known, you can then calculate the total cell number in your suspension (cell number per mL × total media vol (mL)).

Cell viability: The number of living cells is divided by the total number of cells counted (living + dead) and multiplied by 100 (to give a percentage). For example: 352 living cells and 35 dead cells = total cell count of 387. Viability = living cells/total cells * 100 (352/387 * 100) = 90.95%

REFERENCES

[1] Olabarrieta I, L'Azou B, Yuric S, Cambar J, Cajaraville MP. In vitro effects of cadmium on two different animal cell models. Toxicol In Vitro 2001;15(4−5):511−7.
[2] Quinn B, Costello MJ, Dorange G, Wilson JG, Mothersill C. Development of an in vitro culture method for cells and tissues from the zebra mussel (*Dreissena polymorpha*). Cytotechnology 2009;59 (2):121−34.
[3] Klaunig J, Ruch R, Goldblatt P. Trout hepatocyte culture: isolation and primary culture. In Vitro Cell Dev Biol 1985;21(4):221−8.
[4] Seglen PO. Hepatocyte suspensions and cultures as tools in experimental carcinogenesis. J Toxicol Env Health 1979;5(2−3):551−60.

[5] Hansen E. Initiating a cell line from embryos of the snail *Biomphalaria glabrata*. TCA Manual 1979;5 (1):1009–14.

[6] Le Pennec G, Le Pennec M. Induction of glutathione-S-transferases in primary cultured digestive gland acini from the mollusk bivalve *Pecten maximus* (L.): application of a new cellular model in biomonitoring studies. Aquat Toxicol 2003;64(2):131–42.

[7] Chelomin VP, Zakhartsev MV, Kurilenko AV, Belcheva NN. An in vitro study of the effect of reactive oxygen species on subcellular distribution of deposited cadmium in digestive gland of mussel *Crenomytilus grayanus*. Aquat Toxicol 2005;73(2):181–9.

[8] Parolini M, Quinn B, Binelli A, Provini A. Cytotoxicity assessment of four pharmaceutical compounds on the zebra mussel (*Dreissena polymorpha*) haemocytes, gill and digestive gland primary cell cultures. Chemosphere 2011;:91–100.

[9] Parolini M, Binelli A, Cogni D, Provini A. Multi-biomarker approach for the evaluation of the cyto-genotoxicity of paracetamol on the zebra mussel (*Dreissena polymorpha*). Chemosphere 2010;79 (5):489–98.

[10] Canesi L, Ciacci C, Betti M, et al. Rapid effects of 17 beta-estradiol on cell signaling and function of *Mytilus* hemocytes. Gen Comp Endocrinol 2004;136(1):58–71.

[11] Canesi L, Ciacci C, Lorusso LC, et al. Effects of Triclosan on *Mytilus galloprovincialis* hemocyte function and digestive gland enzyme activities: possible modes of action on non target organisms. Comp Biochem Physiol Toxicol Pharmacol 2007;145(3):464–72.

[12] Canesi L, Lorusso LC, Ciacci C, et al. Effects of blood lipid lowering pharmaceuticals (bezafibrate and gemfibrozil) on immune and digestive gland functions of the bivalve mollusc, *Mytilus galloprovincialis*. Chemosphere 2007;69(6):994–1002.

[13] Canesi L, Lorusso LC, Ciacci C, et al. Immunomodulation of *Mytilus* hemocytes by individual estrogenic chemicals and environmentally relevant mixtures of estrogens: in vitro and in vivo studies. Aquat Toxicol 2007;81(1):36–44.

[14] Binelli A, Cogni D, Parolini M, Riva C, Provini A. In vivo experiments for the evaluation of genotoxic and cytotoxic effects of Triclosan in Zebra mussel hemocytes. Aquat Toxicol 2009;91 (3):238–44.

[15] Gómez-Mendikute A, Elizondo M, Venier P, Cajaraville M. Characterization of mussel gill cells in vivo and in vitro. Cell Tissue Res 2005;321(1):131–40.

[16] Parolini M, Binelli A, Cogni D, Riva C, Provini A. An in vitro biomarker approach for the evaluation of the ecotoxicity of non-steroidal anti-inflammatory drugs (NSAIDs). Toxicol In Vitro 2009;23(5):935–42.

[17] Koyama S, Aizawa M. Tissue culture of the deep-sea bivalve *Calyptogena soyoae*. Extremophiles 2000;4(6):385–9.

[18] Kaloyianni M, Stamatiou R, Dailianis S. Zinc and 17β-estradiol induce modifications in Na$^+$/H$^+$ exchanger and pyruvate kinase activity through protein kinase C in isolated mantle/gonad cells of *Mytilus galloprovincialis*. Comp Biochem Physiol Part C: Toxicol Pharmacol 2005;141(3):257–66.

[19] Cornet M. Primary mantle tissue culture from the bivalve mollusc *Mytilus galloprovincialis*: investigations on the growth promoting activity of the serum used for medium supplementation. J Biotechnol 2006;123(1):78–84.

[20] Domart-Coulon I, Auzoux-Bordenave S, Doumenc D, Khalanski M. Cytotoxicity assessment of antibiofouling compounds and by-products in marine bivalve cell cultures. Toxicol In Vitro 2000;14 (3):245–51.

[21] Lambert C, Nicolas J-L, Bultel V. Toxicity to bivalve hemocytes of pathogenic vibrio cytoplasmic extract. J Invertebr Pathol 2001;77(3):165–72.

[22] Mannello F, Canesi L, Gazzanelli G, Gallo G. Biochemical properties of metalloproteinases from the hemolymph of the mussel *Mytilus galloprovincialis* Lam. CompBiochem Physiol B Biochem Mol Biol 2001;128(3):507–15.

[23] Mosmann T. Rapid colorimetric assay for cellular growth and survival: Application to proliferation and cytotoxicity assays. J Immunol Methods 1983;65(1–2):55–63.

[24] Downs CA, Fauth JE, Downs VD, Ostrander GK. In vitro cell-toxicity screening as an alternative animal model for coral toxicology: effects of heat stress, sulfide, rotenone, cyanide, and cuprous oxide on cell viability and mitochondrial function. Ecotoxicology 2010;19(1):171−84.

[25] Fornelli F, Minervini F, Logrieco A. Cytotoxicity of fungal metabolites to lepidopteran (*Spodoptera frugiperda*) cell line (SF-9). J Invertebr Pathol 2004;85(2):74−9.

[26] Cao A, Novás A, Ramos-Martínez JI, Barcia R. Seasonal variations in haemocyte response in the mussel *Mytilus galloprovincialis* Lmk. Aquaculture 2007;263(1−4):310−9.

[27] Strober W. Monitoring cell growth. Current protocols in immunology. New York: John Wiley & Sons, Inc; 2001.

[28] Canesi L, Borghi C, Ciacci C, et al. Short-term effects of environmentally relevant concentrations of EDC mixtures on *Mytilus galloprovincialis* digestive gland. Aquat Toxicol 2008;87(4):272−9.

CHAPTER 4

Measuring Effects at the Gene Transcription Level

Contributed by Chantale André

Chapter Outline

4.1 INTRODUCTION

Ecotoxicogenomics is the application of high-throughput genomic technologies to study the toxicological effects of an environmental chemical or mixtures of interest. The application of these technologies to ecotoxicology can enhance the value of the classical biomarker approach and improve the understanding of toxicity mechanism for previously uncharacterized contaminants. Gene expression and regulation analysis can also be useful to screen compounds for hazard identification, track cellular responses to different doses, and to predict individual variability and sensitivity to toxicants [1]. Reverse transcriptase

Biochemical Ecotoxicology
DOI: http://dx.doi.org/10.1016/B978-0-12-411604-7.00004-0

polymerase chain reaction (RT-PCR) has become the most widely applied technique to study gene expression. This method can easily be used to study the expression of 10 or more genes by using 96- or 384-well microplates. A typical experiment involves the following steps: (1) sampling of the biological material, (2) RNA extraction, (3) reverse transcription of RNA into cDNA, and (4) quantitation of the target cDNA sequence by quantitative PCR (qPCR) instrument. This chapter presents a protocol to analyze the relative gene expression of transcripts involved in toxic stress in rainbow trout hepatocytes using SYBR Green I dye chemistry. It intends to give an overview of the steps and technical aspects needed to generate reliable and quantitative RT-qPCR data (Figure 4.1). Although developed for trout hepatocytes, it should provide enough knowledge to set up an optimized two-step RT-qPCR gene expression experiment and to adapt the method to different species and tissues. Further references providing detailed qPCR-specific information and recommended websites are listed in the reference section.

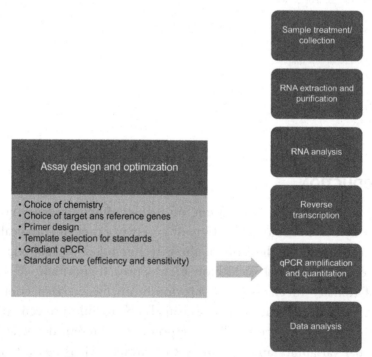

Figure 4.1 Steps involved for gene expression analysis using RT-qPCR.

The principle of the PCR consists of repeating cycles of DNA denaturation, annealing, and elongation steps (Figure 4.2). Template DNA, primers (oligonucleotides of 15–30 mer complementary to the template DNA), and reagents (a cocktail composed of Taq DNA polymerase, nucleotides, magnesium, and buffer) are heated at high temperatures (usually between 95 and 98°C) for 1 min or less followed by a temperature drop for annealing of oligonucleotide primers on the DNA template (usually between 50 and 60°C). At these temperatures, DNA rehybridizes again into double-stranded dimers. Hybridization with the primers also occurs, which consists of complementary oligonucleotide sequences to the template DNA, hence the requirement of knowing the nucleotide sequence of the template DNA. The temperature is then increased at an elongation temperature (usually 70–75°C) to permit the Taq DNA polymerase to synthesize a complementary strand DNA. During that process, the fluorescent dye intercalates in the double-stranded structure of amplified DNA and gives fluorescence. The cycle is then repeated many times.

The PCR methodology is one of the most powerful and widely applied techniques in molecular biology. Theoretically, PCR amplifies target DNA

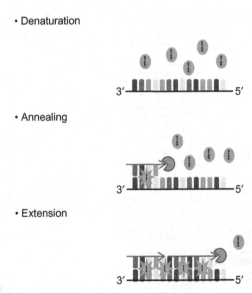

Figure 4.2 Principle of SYBR Green detection in qPCR. The denaturation step contains predominantly single-stranded DNA therefore the signal is only detectable during the annealing and extension steps. As PCR product accumulates, the amount of double-stranded DNA increases and the fluorescence signal increases proportionally.

exponentially, doubling the number of target DNA with each cycle. Cycles are usually repeated 30–40 times giving the potential to produce $\sim 10^9$ molecules of amplified product from one molecule of template. Although this process is designed to be exponential throughout the entire PCR, the real-life behavior of the reaction could be quite different. As the reaction progresses, reagents become limiting and slow down the reaction with a characteristic plateau phase (Figure 4.3). Toward the end of the PCR, DNA amplification is no longer exponential when it reaches a plateau phase, thus making it very difficult to deduce the initial amount of target DNA from the evaluation of amplified DNA at the end of the reaction cycles [2]. Conventional PCR has the disadvantage of relying on end-of-run measurements (gel electrophoresis followed by densitometry analysis) at the plateau phase. The advent of real-time PCR was a significant improvement of the basic PCR technique. The improvement enabled us to follow the kinetics of the reaction in real time and to accurately quantify the starting amount of nucleic acid during the exponential phase of the PCR reaction. The method is based on the use of a reporter dye whose fluorescence increases as the amount of DNA increases at each successive

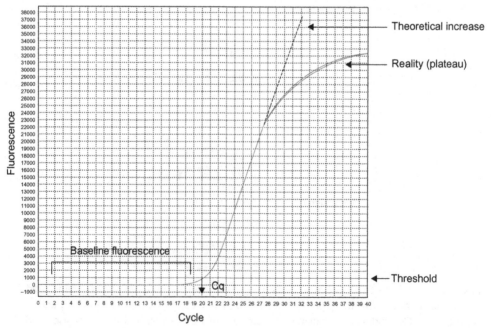

Figure 4.3 Typical amplification plot and commonly used data terms.

amplification cycle (Figure 4.3). At the beginning of the reaction, the signal is under the detection limit of the fluorescent dye even though product accumulates exponentially. This background signal in all wells is used to determine the baseline fluorescence (blank). As the reaction progresses, enough amplified product accumulates until it rises above the background fluorescence signal, which eventually levels off and plateaus. The relative fluorescence units that yield a statistically significant fluorescence increase over the calculated baseline signal is called the threshold [3]. The cycle number at which the signal becomes greater than the threshold is called the quantification cycle (Cq). The greater the initial amount of template in the sample, the earlier the Cq value for that sample will be reached (e.g., increase in fluorescence at the 9th cycle instead of the 20th cycle). The Cq is the basic metric of qPCR and is an essential component in producing accurate and reproducible data.

4.2 MATERIAL

4.2.1 Reagents

Oligonucleotides: Custom nucleic acids are available from numerous suppliers such as IDT (Coralville, IA). Usually the oligonucleotides do not contain too many G or C nucleotides as they increase the denaturation temperature. The oligonucleotides are usually between 12 and 30 nucleotides long, which forms the basis of specificity of the method. The oligonucleotide sequences should not have complementary sequences with each other as they will form heat-stable primer dimers.

Commercial RNA extraction kit: Any commercial kit could be used, otherwise a crude RNA fraction could be obtained from detergent/protease-treated and phenol-chloroform iso-amyl alcohol extraction of the sample (see the method in Chapter 10) followed by DNase treatment to remove DNA. For fish tissues, the Qiagen RNeasy mini kit worked well.

Nuclease-free water: Consists of diethylpyrocarbonate (DEPC)–treated water, which is also commercially available at many suppliers (Ambion 9915G).

Commercial qPCR Mastermix or core components: KCl, buffer, pH 8.4, dNTPs, Hot Start enzyme, $MgCl_2$, and SYBR Green I.

cDNA synthesis: Commercial two-step RT kit (e.g., QuantiTect, Qiagen).

TE 1× buffer, pH 7.5−8: 10 mM Tris, 1 mM EDTA (adjust pH with HCl before addition of EDTA). Molecular biology grade ethanol (70 and 95%).

4.2.2 Equipment

Primer design software and tools: Homology sequence database such as Primer-BLAST uses Primer3 to design PCR primers to a sequence template.

Real-time PCR thermal cycler, $-85°C$ freezer, filter barrier pipette tips, nuclease-free and pyrogen-safe (heat-stable) microtubes (e.g., Axygen MCT-150-A), qPCR 96-well microplates, optical adhesive sealing film, microcentrifuge (1.5–2 mL microtubes), and microplate mixer (optional).

Specialty analysis software: GenEx Enterprise software, MultiD Analyses AB.

Automated microfluidic-based system: Experion Automated Electrophoresis System (Bio-Rad), using the Experion RNA StdSens Analysis kit (Bio-Rad), or Agilent 2100 Bioanalyzer, or Agarose gel electrophoresis apparatus and densitometry software.

Spectrophotometer: NanoDrop ND-1000 spectrophotometer (or spectrophotometer with quartz cuvette for UV scanning).

4.3 PROCEDURE

4.3.1 Choice of Detection Chemistry

There are several fluorescent detection chemistries to choose from to monitor amplification in qPCR. They can be largely classified into two categories: nonspecific DNA binding dyes, and sequence-specific oligoprobes. SYBR Green I is the most widely used nonspecific DNA binding dye. It binds all double-stranded DNA molecules (including nonspecific reaction products, e.g., primer dimers) and emits a fluorescent signal (494 nm excitation and 521 nm emission maxima) upon binding, allowing detection in the extension step of the real-time PCR (Figure 4.1). A significant disadvantage of DNA-binding dye assays is that the specificity of the reaction is determined entirely by the primers [4]. A well-optimized reaction combined with melting curve analysis is therefore essential to ensure specificity and accurate results. Sequence-specific probes, on the other hand, are significantly more costly but enable multiplex qPCR (amplification and detection of two or more different amplicons in the same reaction). This RT-qPCR protocol uses the double-strand DNA binding dye SYBR Green I as it is the most cost-effective chemistry and ideal for primer validation, even if you are planning on using a probe-based chemistry later.

4.3.2 Primer Design

qPCR assays for various targets can be taken from the literature or in databases depending on the species. However, it is primordial to validate these primers in your laboratory and evaluate the amplification efficiency and the assay's sensitivity. The selection of primers for a qPCR reaction will determine the efficiency and specificity of the PCR.

4.3.3 Selecting the Target Nucleotide Sequence (Target and Reference Genes)

RT-qPCR data are most commonly normalized using reference genes. It is recommended to test many reference genes (>3) for stability of expression as they sometimes change with the treatment in addition to "biomass" for normalization. A good reference gene does not exhibit variation in expression across different sample treatments or time points [5].

There are often several available sequences for the same gene since some genes have been sequenced numerous times. Polymorphism, mutation, or error can occur resulting in publication of different sequences. BLASTn search (see http://blast.ncbi.nlm.nih.gov) your sequence of interest to align all matched sequence and reveal mismatches. If any are found, avoid this region for primer design.

Input the target nucleotide sequence (or the primers taken from the literature) into Primer-BLAST (see http://www.ncbi.nlm.nih.gov/tools/primer-blast/). Adjust the PCR amplicon size parameter to 70–200. Specify a database and an organism for database search if you are only amplifying cDNA from a specific organism. Searching all organisms will be much slower and off-target priming from other organisms is irrelevant.

Enter the Primer-BLAST suggested primer pairs sequences into NetPrimer (PREMIER Biosoft, Palo Alto, CA; http://www.premierbiosoft.com/netprimer/) and analyze for secondary structures (hairpins, self-dimers, and cross-dimers). Select primer pairs with the higher rating (predicted amplification efficiency), a stable 5' end (lower ΔG), and an unstable 3' end (less negative 3' end stability value). Try to avoid secondary structures with binding energy < -5 kcal/mol.

Copy the amplicon sequence (region amplified by the primer pairs you have chosen) and enter sequence in mfold (see http://www.bioinfo.rpi.edu/applications/mfold) to analyze for secondary structures that could prevent efficient amplification.

Adjust the folding temperature (use predicted annealing temperature) and the ionic conditions. For the purposes of folding, Na^+ is considered equivalent to K^+ and Mg^{2+} is equivalent to Ca^{2+} [6].

Select two to three primer pairs to be evaluated. Order desalted purified primers from an oligonucleotide design company. Our primers were manufactured by IDT (Coralville, IA) using the Lab ready service.

Upon arrival, resuspend primers in DNase/RNase-free water or TE buffer (oligos will be more stable) following the manufacturer's instructions. Ready-to-use stock solution of primers is also available at extra cost (Lab ready service). Prepare aliquots at 100 μM (10 × working concentration) and store at −20°C.

4.3.4 Preparation of Nucleic acid Template

Unlike DNA, RNA is extremely delicate, because ribonucleases (RNases) are ubiquitous, extremely stable, and active. RNA can be easily degraded by RNases during sample harvesting, RNA extraction, purification, and storage. Proper handling of tissues and cells prior to RNA isolation is therefore crucial for obtaining high-quality RNA and great precautions must be taken to avoid RNase contamination of the RNA samples to prevent degradation. RNA preparations are often contaminated by genomic DNA (gDNA). Even an infinite amount of gDNA will most probably be amplified and result in erroneous or unreliable qPCR data [7]. The selected RNA extraction procedure should therefore include a DNase I treatment step.

A summary of precautions that should be taken when working with RNA is as follows:

- Store samples in appropriate conditions (−80°C) and/or in RNA stabilization agent, e.g., RNA *later*.
- Wear clean gloves and lab coat at all times (skin is a major source of RNases).
- Use RNase-free plasticware and filter barrier tips.
- Change tips between every liquid transfer to avoid cross-contamination.
- Minimize handling time (process samples in small batches).
- Include a DNase I treatment to remove any possible contamination of gDNA.

There are many methods available for RNA isolation and the most widely used can be divided into two groups: phenol-based extraction (e.g., phenol/chloroform) and filter-based isolation (e.g., silica spin columns). The phenol-based extraction usually yields a high level of RNA but has the disadvantage of using hazardous reagents and can result in more carryover of compounds that can interfere with downstream processing. Column-based methods are less time-consuming and the RNA yields and purity are generally very good, but the main disadvantage is the higher cost. Many extraction kits are commercially available from numerous suppliers (e.g., Qiagen, Bio-Rad, Ambion, Promega, Stratagene, Invitrogen, and Sigma-Aldrich) for various sample types. Commercial kits are reliable, provide uniform performance, and are less time-consuming. They are relatively more expensive than homemade reagents, but if you can afford it they are a good alternative. Further references are provided that present detailed RNA isolation techniques for a variety of sample types and could be helpful for appropriate isolation technique selection if you are not using a commercial kit [8,9].

4.3.4.1 RNA Extraction from Cultured Cells

After the exposure period, completely aspirate the cell culture/exposure medium and disrupt the hepatocytes (500,000 cells) by adding 350 µL of lysis buffer (supplied in kits) directly to the wells. Mix with the pipette. Alternatively (e.g., if you need to keep some of the cells for another application) completely aspirate the cell culture medium, suspend cells (500,000 cells) in 500 µL of ice-cold PBS, and centrifuge at $400 \times g$ for 5 min. Aspirate supernatant, add 350 µL of lysis buffer to the cell pellet, and vortex to mix.

To prevent unwanted alteration of gene expression and RNA degradation, work as quickly as possible and add lysis buffer directly to the wells when possible. Otherwise, minimize time spent in ice-cold PBS prior to homogenization and RNA stabilization.

Pipette the entire lysate into a QIAshredder spin column (Qiagen) placed in a 2 mL collection tube and centrifuge at maximum speed ($16,000 \times g$) for 2 min at room temperature.

At this point the lysates can be safely stored at $-80°C$ for months until RNA extraction.

Extract the RNA using the RNeasy Plus mini kit (or any appropriate commercial RNA extraction kit) following the manufacturer's instructions.

If not using a kit including a gDNA removal step, consider adding a DNase I treatment after this step.

4.3.4.2 Assessment of RNA Quality

RNA concentration can be determined from the absorbance at 260 nm (A260) with a UV-visible spectrophotometer using a quartz cuvette or the NanoDrop spectrophotometer. RNA can also be quantified fluorometrically using the fluorescent nucleic acid dye RiboGreen whose fluorescence increases upon binding to RNA. The fluorescent method is more sensitive (1−200 ng range) and allows quantification in a 96-well format. The NanoDrop is actually the most widely used method as it allows us to quickly assess RNA concentration and purity in a very small sample (1−2 μL) by measuring the A260 nm/280 nm ratio (e.g., mostly protein contamination). In addition, the A260/A230 ratio can be used to evaluate extraction component carryover (e.g., phenol, guanidine isothiocyanate, EDTA).

Some RNA quantification guidelines are given to the analyst:

- Choose an RNA quantification method and stick to it. Do not compare concentrations obtained from different methods as they often generate different results.
- A260 readings should be between 0.15 and 1.0 to generate accurate results. If the reading is higher than 1.0 you should dilute your sample, preferably in neutral RNase-free TE buffer.
- Use a blank with the same solution you used to suspend or elute the RNA (water or TE 1×). A280 decreases with increasing pH. Measurement at neutral pH is recommended for purity assessment.
- Accurate pipetting is crucial, especially when loading a very small volume. With the NanoDrop system, using a volume of 2 μL usually gives more reproducible results.
- Conversion of A260 to RNA concentration: 1.0 unit A260 = 40 μg/mL RNA.
- A260/A280 ratio should be approximately 1.8−2.0.
- A260/A230 ratio should be approximately 2.0−2.2.
- Exclude samples that have appreciably lower ratio values (use samples with similar ratio values). Some impurities can be removed by extraction of the sample with n-butanol [9].

4.3.4.3 RNA Integrity Analysis

RNA integrity is a critical element in a gene expression analysis. Partially degraded samples can compromise the success of an RT-qPCR experiment.

Furthermore, differences in RNA quality between samples can lead to misinterpretation of gene expression (always use samples with similar quality). Automated microfluidic-based systems like the Experion (Bio-Rad) automated electrophoresis station and the Agilent Bioanalyzer have become the method of choice for RNA integrity assessment allowing rapid RNA quality determination and quantity estimation using a very small volume of sample (1−4 μL) on a chip. Both instruments allow a visual inspection of RNA integrity, generate approximated ratios between the mass of ribosomal subunits (28 S/18 S), and use a numerical system that provides an objective and reliable assessment of the RNA integrity [10]. The Experion (Figure 4.4) calculates an RNA Quality Index and the Bioanalyzer assigns RNA Integrity Numbers. Carefully follow the manufacturer's instructions provided with the electrophoretic chips. If you do not have access to a microfluidic system, the conventional method to assess RNA integrity is denaturing agarose gel electrophoresis in the presence of a fluorescent dye. Formaldehyde (5−10%) denatures the secondary structure of RNA, which could alter migration patterns and provide gels with sharper bands that are easier to interpret. The intensity of the 28 S and 18 S band can be quantified using a scanner and densitometry analysis software. Intact total RNA should have sharp 28 S and 18 S bands, and the ratio of 28 S rRNA to 18 S rRNA should be approximately 2:1 (ratio between 1 and 2 usually indicates intact RNA, depending on species and tissues). However, 28 S rRNA is more susceptible to degradation and thus serves as a good sentinel molecule. This method is inexpensive but is more time-consuming and requires a larger amount of sample. Protocol for formaldehyde agarose gel electrophoresis can be found in Bustin [7].

To prevent RNA degradation after quality assessment, keep RNA on ice and reverse transcribe to a much more stable cDNA immediately after integrity analysis.

4.3.4.4 RT

RT of RNA to cDNA can be carried out using either a one-step or a two-step method. The specific reaction conditions vary upon the particular protocol and commercial kit used, but they all contain the same basic components: RNA samples to be converted, dNTPs, primers, buffer, RNase inhibitors, and a reverse transcriptase enzyme. The one-step method requires a single reaction mix since the RT and the qPCR amplification occur in the same tube.

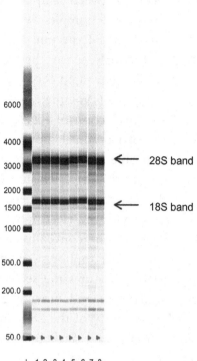

Well ID	Ratio [28S/18S]	RQI	RQI Classification
Ladder			
1	1.23	9.6	Green
2	1.26	9.6	Green
3	1.09	9.5	Green
4	1.46	9.8	Green
5	1.41	9.7	Green
6	1.16	9.5	Green
7	1.32	9.5	Green
8	1.36	9.6	Green

Figure 4.4 Example of Experion automated electrophoresis system gel image and associated RNA quality results.

Although it has the advantage of being faster and more convenient, it requires sequence-specific RT primer for cDNA synthesis and limits the assay optimization. If you are interested in making a large amount of cDNA to use for multiple assays, the two-step qPCR is recommended. It makes it possible to use random, oligo-dT or gene-specific primers, to optimize the qPCR assay separately to gain sensitivity, and to store cDNA for later use. Table 4.1 describes the characteristics of the different primer types that can be used for RT. Many commercial two-step RT kits are available (Applied Biosystems, Bio-Rad, Invitrogen, Qiagen, and Stratagene). When choosing a cDNA synthesis kit, keep in mind the size of the transcribed fragment and the downstream applications you are interested in. A nonexhaustive list of kits can be found   at   http://www.biocompare.com/Cloning-and-Expression/10951-cDNA-Synthesis-Kits/.

Prepare the RT reaction according to the supplier's instructions and apply the following recommendations:

- Keep the RNA samples on ice.
- Use equal starting amounts of RNA for all RT reactions.
- When possible, perform all RT reactions at the same time or the same day.
- If proceeding samples are in different batches, include a positive control in all batches.
- Include a no-RT control in all batches to confirm the absence of gDNA.
  - The no-RT control reaction contains all the RT reaction components except the reverse transcriptase. Replace the enzyme by an equivalent volume of RNase-free water.

Table 4.1 RT Primer Types for cDNA Synthesis

RT Primers	Features
Random primers	Enables RT from the entire RNA population (mRNA, rRNA, tRNA, and so forth)
	Can be used when the template has extensive secondary structure:
	Produces the greatest yield (but favors rRNA)
	Low abundance messages may be under-represented
	Leads to short, partial-length cDNAs
	cDNA produced can be used for many different targets
Oligo-dT primers	Selectively reverse transcribes mRNAs starting from the poly-A tail at the 3′end
	Limits the amount of rRNA (e.g., will not reverse transcribe 18 S rRNA)
	5′ end of the mRNA may be under-represented
	May not efficiently reverse transcribe partially degraded RNA (if loss of intact poly-A tail)
	cDNA produced can be used for many different targets
Gene-specific primers	Primers allows RT of a specific target
	cDNA produced cannot be used for assaying other genes

- Prepare a pooled cDNA-positive control for assay validation: remove an equal volume of the cDNA samples (treated and nontreated samples) and place in a 1.5 mL RNase/DNase/pyrogen-safe microtube.
- Prepare a 4 ng/μL dilution (for standard curve) and a 1 ng/μL dilution of the pooled cDNA-positive control using RNase-free water.
  - Concentration is based on initial RNA amount in RT.
- Prepare a working dilution of your sample's cDNA using RNase-free water so that ±5 μL will be added to each qPCR reaction (e.g., 1 ng/μL).
  - To prevent qPCR inhibition, no more than 10% of the final PCR reaction volume should derive from the RT reaction. Diluting the template will allow you to pipette a larger volume (e.g., 5 μL) thus reducing the risk of pipetting errors that often occur when pipetting very low volumes.
- Aliquot the cDNA samples (≈10 μL) and store at −20°C (−80°C for longer storage).

4.3.5 Assay Optimization

4.3.5.1 Primer Validation

For each primer pair, test the specific amplification of the template using a single primer concentration (e.g., 300 nM), 5−10 ng of cDNA, and the same cycling conditions. If for a given target none of the primer pair yields satisfying results, an additional optimization experiment can be run (gradient annealing temperatures and primer concentrations 100−300 nM) to determine alternative assay conditions for those targets. For experiments studying multiple genes, it is recommended (and much easier) to find an annealing temperature (e.g., 60°C) suitable for all primer pairs thus enabling us to study them all in the same run. To prevent contamination issues, maintain a separate area for PCR setup. Never open an amplified PCR product tube or plate in the PCR setup area and use a dedicated set of equipment and supplies (e.g., pipettes, filtered barrier tip). The use of a flow hood or benchtop enclosure with UV irradiation can be useful to prevent contamination. However, all plasticware (microtubes, tips, and microplates) must be removed from the enclosure prior to UV irradiation as their irradiation can lead to PCR amplification inhibition. Prepare your plate layout before you start to avoid doubts and errors while pipetting.

Thaw on ice by using 10× reverse and forward primers, aliquot of pooled cDNA-positive control (1 ng/μL), and commercial SYBR Green Mastermix (containing optimized reaction buffer, dNTPs, $MgCl_2$, SYBR Green, and Taq polymerase).

Commercial mastermixes are widely used, well optimized, and convenient. You can also choose to prepare your own SYBR Green Mastermixes by manual addition of all the reaction components. Although it increases the risk of variability due to pipetting errors, it allows the flexibility of adjusting the component's concentration (e.g., $MgCl_2$) according to experimental needs. Core kits containing all reaction components in separate tubes are also commercially available.

In sterile 1.5 mL RNase/DNase /pyrogen-safe microtubes:

- Prepare a working dilution of all forward and reverse primers: Dilute $10\times$ stock of primers 1/10 in RNase-free buffer. Vortex to mix and keep on ice.

- Prepare a qPCR reaction mastermix for each primer pair to assay: Pipette in the following order: the RNase-free water, the SYBR Green Mastermix, and the forward and reverse primers. Adjust the volumes of each component to the number of reactions you need to run (Table 4.2). Run each reaction in duplicate. Gently pipette up and down to mix and keep on ice.

- SYBR Green is light sensitive, so use aluminum foil to protect it from light. Using a Hot Start enzyme, you can set up your PCR plate at room temperature.

- Transfer 20 μL of each qPCR reaction mastermix (for a 25 μL reaction volume) to the bottom of the appropriate wells of a 96-well qPCR plate (alternatively you can use qPCR strips or qPCR tubes, depending on your instrument). Include duplicate "no template" control wells (NTC) for each primer pair. Avoid bubbles.

Table 4.2 Example of qPCR Reaction Mastermix

Component	Volume Per 25 μL Reaction	Final Concentration
$2\times$ SYBR Green Mastermix	12.5	$1\times$
10 μM forward primer	0.75	300 nM
10 μM reverse primer	0.75	300 nM
RNase-free water	6	
TOTAL	20 μL	

Using the $2\times$ SYBR IQ Green Supermix (Bio-Rad) the $1\times$ concentration corresponds to 50 mM KCl, 20 mM Tris-HCl, pH 8.4, 0.2 mM of each dNTP, 25 units/mL iTaq DNA polymerase, 3 mM $MgCl_2$, SYBR Green I, and 10 nM fluorescein. For each component, multiply the volume needed for one reaction by the number of reaction. For each primer pair to test include: 2 reaction wells + 2 NTC wells = 4 reactions. Prepare a final mastermix volume 10% greater than the calculated required volume for the total number of reactions.

- Most qPCRs are performed in $10-25 \mu L$ reaction volumes. Keep the reaction volumes constant across all samples that need to be compared.
- Add $5 \mu L$ of diluted cDNA to the appropriate wells (all reaction wells). Add $5 \mu L$ of RNase-free water to the NTC wells
- To prevent qPCR inhibition, no more than 10% of the final reaction volume should derive from the RT reaction.
- Seal the plate with optically clear film to prevent evaporation.
- Important: Ensure the plate is tightly sealed to prevent evaporation. Avoid fingerprints on top of the film.
- Optional step: To ensure that all components are at the bottom of the wells and that there are no bubbles, centrifuge the plate at $1000 \times g$ for 30 s and then mix 30 s at 1500 rpm on a PCR plate mixer (e.g., Eppendorf's Mixmate plate mixer).
- Load the plate on the instrument and run the cycling protocol. Include a melt curve analysis at the end of the program.
- Adapt the cycling parameters to the SYBR Green Mastermix being used to adequately activate the enzyme. Refer to the mastermix (or enzyme) supplier's instructions.
- Following amplification, analyze the melt curves. For each target, select the primer pairs that have a melt curve profile showing a single melt peek in the sample wells and no amplification in the NTC (Figure 4.5). Choose the primer pair with the lower Cq (preferably below 30) and proceed to amplicon size analysis.
- Do not open the PCR plate/tube containing amplicons (amplified DNA) in the PCR setup area. Move to another room or area. It is very difficult to get rid of contamination from amplified DNA.
- Confirm that the selected primer pair yields an amplicon of the expected size. Run at least one PCR product for each selected primer pair either on an automated microfluidic system (e.g., Experion system using the DNA 1K analysis kit) or on a SYBR Green-stained 2% agarose gel.
- If available, work under a laminar flow hood and open and close all tubes/plates carefully to avoid splashing. Following analysis, clean workbench with 70% ethanol and irradiate overnight in UV light.
- Select the optimal primer pair for each target. It should be a primer pair that yields a single product of the expected size, with the lowest Cq and absence of amplification in the NTC.

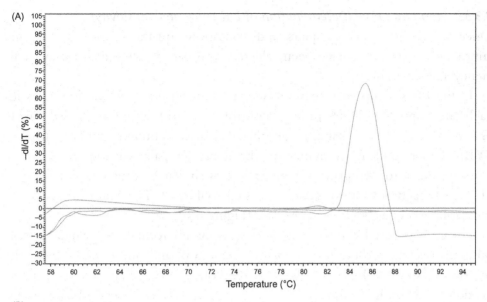

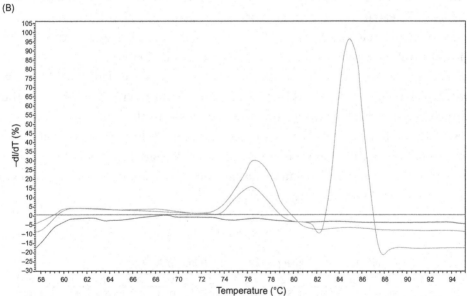

Figure 4.5 Melt curve analysis representing specific amplification and primer dimers. A single peak in the sample melt curve and no peak in the NTC sample indicate a single PCR product (A). Additional peaks of lower melting temperature or a peak in the NTC sample indicate primer dimers (B).

4.3.5.2 Standard Curve (Determination of Efficiency and Sensitivity)

Once the optimal primer pairs and conditions (primer concentration and annealing temperature) have been selected, you can determine the assay's efficiency and sensitivity.

In the PCR setup area, thaw on ice: $10 \times$ reverse and forward primers for each target, pooled cDNA-positive control (4 ng/μL), and commercial SYBR Green Mastermix (containing optimized reaction buffer, dNTPs, $MgCl_2$, SYBR Green, and Taq polymerase). Each primer pair will require 20 μL of cDNA control (or template of your choice as similar to your samples as possible), 2–4 reaction wells for standards and 3 wells for NTC. You can assay three standard curves per 96-well microplate (Figure 4.6).

For each selected primer pair prepare eight microtubes for standard serial dilution, two microtubes for the primer's dilution, and one microtube for the reaction mastermix. Prepare a working dilution of forward and reverse primers as described earlier. Also prepare the mastermix as described previously. Calculate 27 reactions per primer pair and prepare a final volume at least 10% greater than the calculated required volume. For a 25 μL reaction qPCR, you will need to prepare 600 μL of mastermix per standard curve.

Prepare a fivefold dilution series (8 points) of your pooled cDNA-positive control (Table 4.3). Add the RNase-free water in all microtubes first and then add the cDNA, vortex gently to mix, and continue to the next tube.

A 3- to 10-fold dilution series can be prepared. Select an appropriate dilution to end up with a range as broad as possible—at least five valid dilution points—so that the standard curve range covers the anticipated Cq values of your experimental samples (Figure 4.7).

	Primer pair #1			Primer pair #2			Primer pair #3					
	1	2	3	4	5	6	7	8	9	10	11	12
A	STD 1	STD 1	STD 1	STD 1	STD 1	STD 1	STD 1	STD 1	STD 1	NTC	NTC	NTC
B	STD 2	STD 2	STD 2	STD 2	STD 2	STD 2	STD 2	STD 2	STD 2	NTC	NTC	NTC
C	STD 3	STD 3	STD 3	STD 3	STD 3	STD 3	STD 3	STD 3	STD 3	NTC	NTC	NTC
D	STD 4	STD 4	STD 4	STD 4	STD 4	STD 4	STD 4	STD 4	STD 4			
E	STD 5	STD 5	STD 5	STD 5	STD 5	STD 5	STD 5	STD 5	STD 5			
F	STD 6	STD 6	STD 6	STD 6	STD 6	STD 6	STD 6	STD 6	STD 6			
G	STD 7	STD 7	STD 7	STD 7	STD 7	STD 7	STD 7	STD 7	STD 7			
H	STD 8	STD 8	STD 8	STD 8	STD 8	STD 8	STD 8	STD 8	STD 8			

Figure 4.6 Example of standard curve microplate setup.

Add 20 µL of reaction mastermix to their respective reaction wells (standards and NTC, refer to Figure 4.6). Avoid bubbles. Add 5 µL of each cDNA dilution (in triplicate) to its respective well, 5 µL of RNase-free water to the NTC wells, and seal the plate with optically clear film and perform the qPCR steps as explained previously. At the end of the reaction adjust the instrument software analysis settings:

A. In linear view, look at the amplification plots and examine the baseline settings (baseline default setting is often 3–15). Ensure that the baseline is set to include the background fluorescence present in the initial cycles and to end at least two cycles before the earliest sample that emerges

Table 4.3 Preparation of a Fivefold Dilution Series for Standard Curve Generation

Microtube	Volume (µL) cDNA	Volume (µL) RNase-Free Water	ng cDNA/5 µL
S1 (4 ng/µL cDNA control)	5	0	20
Fivefold serial dilutions:			
S2	5 (of S1)	20	4
S3	5 (of S2)	20	0.8
S4	5 (of S3)	20	0.16
S5	5 (of S4)	20	0.032
S6	5 (of S5)	20	0.0064
S7	5 (of S6)	20	0.00128
S8	5 (of S7)	20	0.000256

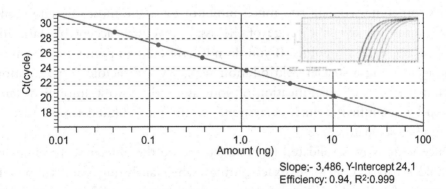

Slope;- 3,486, Y-Intercept 24,1
Efficiency: 0.94, R^2:0.999

Figure 4.7 Example of standard curve obtained from amplification of a threefold dilution series of rainbow trout hepatocyte cDNA.

above background (in this case, your highest concentrated standard). Adjust manually if necessary. Verify that the data in the baseline have been set to zero.

B. In log view, ensure that the threshold is set approximately in the middle of the exponential amplification region (linear portion of the plot in log view) across all of your sample amplification plots. Adjust manually if necessary.

Threshold and baselines are automatically set by the instrument software using default settings. Although they are often appropriate, it is important to verify those settings and to adjust them manually if necessary to generate good data.

Verify that the amplification plot has a normal shape (Figure 4.3) for each standard. Examine replicates and ensure that the difference between the replicates is <0.5 Cq. Remove abnormal samples from analysis.

Examine melting curve profiles for every sample dilution to ensure they yield a single peak at the anticipated T_m and check for the absence of amplification in the NTC.

qPCR instrument software automatically generates standard curve (by plotting the Cqs of each standard against the known cDNA quantity) and calculates slope, R^2, and reaction efficiency (efficiency $= 10^{(-1/\text{slope})} - 1$). Validate that your results fall in the following range:

- Slope between -3.2 and -3.5
- $R^2 > 0.98$
- Reaction efficiency 90–110% (as close to 100% as possible).

Remove points at both ends of the curve if necessary (very concentrated samples sometimes contain too much inhibitor or as the most diluted samples, might exceed the dynamic range of the assay) to obtain acceptable efficiency and R^2. Keep at least 5 points. If you do not have enough points run a standard curve again using a smaller fold dilution (e.g., try a threefold serial dilution). Establish the limit of quantification of your assay (the lowest amount of target that can be quantified), which corresponds to the most diluted sample that you kept in your standard curve.

Once your assay is validated for a target, record the efficiency, baseline, and threshold settings. Use these exact settings when analyzing your samples for that target. You should also run a new standard curve to validate your reaction efficiency each time you order a new batch of primers.

	1	2	3	4	5	6	7	8	9	10	11	12
A	#1	#2	#3	#4	#5	#6	#7	#8	#9	#10	#11	#12
B	#1	#2	#3	#4	#5	#6	#7	#8	#9	#10	#11	#12
C	#13	#14	#15	#16	#17	#18	#19	#20	#21	#22	#23	#24
D	#13	#14	#15	#16	#17	#18	#19	#20	#21	#22	#23	#24
E	#25	#26	#27	#28	#29	#30	#31	#32	#33	#34	#35	#36
F	#25	#26	#27	#28	#29	#30	#31	#32	#33	#34	#35	#36
G	#37	#38	#39	#40	#41	#42	#43	#44	#45	Pos C	No-RT	NTC
H	#37	#38	#39	#40	#41	#42	#43	#44	#45	Pos C	No-RT	NTC

Figure 4.8 Example of microplate setup for analysis of experimental samples.

4.3.5.3 Determination of Gene Transcripts by qPCR

Thaw on ice $10 \times$ reverse and forward primers, commercial SYBR Green Mastermix, pooled positive control cDNA, no-RT control, and experimental samples. A maximum of 45 experimental samples can be analyzed on a 96-well plate (Figure 4.8).

If for a given target, all experimental samples cannot be analyzed on the same plate, cDNA–positive control can be used as an interplate calibrator.

Prepare primer dilution and mastermix as described previously. The total volume should include enough for all reaction wells + 10% extra volume. Transfer 20 µL of reaction mastermix to each well. Add 5 µL of template (samples, positive control, and no–RT control) to their respective wells. Add 5 µL of RNase-free water to the NTC wells. Seal plate, load the plate in the thermal cycler, and run the appropriate cycling protocol as determined earlier.

4.4 DATA ANALYSIS-RELATIVE QUANTIFICATION WITH REFERENCE GENE NORMALIZATION

Relative quantification is the more common method of gene expression analysis and is expressed as the fold change in gene expression relative to an untreated control normalized with one or multiple reference genes. Adjust your baseline and threshold settings as determined in your standard curve experiment. Confirm in controls for no amplification in the NTC, no amplification in the no-RT control, the absence of gDNA, and amplification of positive control. Examine melt curve profiles as described previously. Determine the replicates' responses and ensure that the difference between each replicate is <0.5 Cq. Remove abnormal samples from analysis.

At this point amplification in the NTC indicates contamination. For a given target, Cq of positive control should not vary more than one cycle from one run to another.

Determine Cq for all samples and verify they are within the dynamic range of the assay. Do not extrapolate outside the range of the standard curve.

Import Cq values to a spreadsheet or to a qPCR data analysis software like GenEx (MultiD) or qBase (Biogazelle).

The MIQE guideline recommends the use of a specialist data analysis software [11]. This software speeds up data analysis and usually allows us to identify and deal with missing data and outliers, apply efficiency corrections, interplate variation corrections using the interplate calibrator, average technical repeats, and so forth. It also provides integrated algorithms for best reference gene selection, normalization to one or more reference genes, and calculates relative gene expression.

For reference gene validation, select the best reference genes (genes with the most stable expression throughout all samples) using tools like geNorm [12] or NormFinder [13]. Expression levels should not change with treatment and replicates. Cq variation should be ≤ 0.5. Normalize with the selected reference gene(s) and average technical replicates. The treatment groups could be then normalized against the control (no treatment or reference). Determine fold change in expression using the classical $\Delta\Delta Cq$ method:

$$\text{Fold change} = 2^{-\Delta\Delta Cq}$$

where:

$\Delta Cq = Cq \text{ target gene} - Cq \text{ reference gene}$

$\Delta\Delta Cq = \Delta Cq \text{ sample} - \Delta Cq \text{ untreated control}.$

To be accurate, the $\Delta\Delta Cq$ method requires identical or very similar reference gene and target gene efficiencies ($<10\%$ difference). If the efficiency difference is more than 10% (and you cannot redesign the assay) then an efficiency correction method must be used as described in [14]. A troubleshooting guide is provided to assist the laboratory investigator in optimizing and working through RT-qPCR procedures (see Table 4.4).

Table 4.4 Troubleshooting Guide for Performing qPCR Experiments for Gene Expression

Problems	Possible Cause	Comments and Suggestions
Low RNA yield	Too much starting material	Reduce the amount of starting material.
	Insufficient disruption and homogenization	Increase lysis buffer volume, reduce amount of starting material, and increase homogenization time.
	Incomplete removal of cell culture medium	Ensure you completely remove culture medium.
Low A260/A280 ratio	pH of solution is acidic	Dissolve sample in TE instead of water.
	Absorbance is outside the linear range	Dilute sample to bring absorbance into linear range.
RNA degraded	Inappropriate handling of starting material	Ensure that samples are properly stabilized and stored. Snap freeze in liquid nitrogen or stabilize in RNA later immediately after harvesting. Store at $-80°C$. For cell culture samples, minimize washing steps. Perform the extraction procedure quickly.
	RNase contamination	Use RNase-free consumables and reagents. Be careful not to introduce RNases coming from the surface of the skin or from laboratory equipment, e.g. change gloves frequently (every time you touch a potentially contaminated surface), keep tubes closed whenever possible.
No amplification (or very high Cq)	Pipetting error/missing reagent during qPCR setup	Check concentration and storage conditions of reagents and repeat experiment.
	Wrong detection step	Ensure that fluorescence detection is activated and takes place in the extension step.
	Enzyme not activated	Ensure that the appropriate initial activation at $95°C$ was performed.
	Inappropriate annealing temperature	Have the optimal annealing temperature (perform an annealing temperature gradient ($2°C$ increment)).
	Insufficient annealing/extension times	Verify the amplicon length on agarose gel. Increase extension time if necessary (up to 30s).
	$MgCl_2$ concentration not optimal	Increase (up to 6 mM) $MgCl_2$ concentration in 0.5 mM increments to find optimal concentration.
	Template contains inhibitors	Dilute template (1:10−1:100) and repeat the assay.
	Poor primer design	Try another primer pair.

(Continued)

Table 4.4 Troubleshooting Guide for Performing qPCR Experiments for Gene Expression—(cont.)

Problems	Possible Cause	Comments and Suggestions
Amplification in NTC	Primer dimers	Examine melt curve to confirm the presence of primer dimers. If primer dimers are confirmed then increase annealing temperature, decreasing primer concentration and redesign assay with new primer pair.
	Contamination of reagents	Discard reaction components and repeat with new reagents (or new aliquots).
	Contamination during plate setup	Repeat using additional precautions. Use filter barrier tips and change tips between all samples and reagents.
Amplification in no-RT control	gDNA contamination	Add a DNase I treatment to RNA sample. Design primers to span exon–exon boundaries
Poor linearity of the cDNA dilution series in standard curve ($R^2 < 0.98$)	Template amount too high/too low	Remove points at the extremity of the curve, decrease/increase starting amount, and repeat dilution series.
	Reaction components not properly mixed	Repeat serial dilution and ensure all reaction components are properly mixed and serial dilutions are vortexed (15s).
Multiple peaks in melt curve	Primer dimers and nonspecific amplification	Increase annealing temperature, lower $MgCl_2$ concentration, and redesign assay with new primer pair.
	AT-rich subdomain	Shoulder in the melt curve can indicate an AT-rich subdomain rather than nonspecific amplification. Run amplicon on agarose gel to check specificity
High variability between replicates	Poor pipetting Evaporation	

REFERENCES

[1] National Research Council (US). Committee on applications of toxicogenomic technologies to predictive toxicology. Applications of toxicogenomic technologies to predictive toxicology and risk assessment. Washington DC: National Academies Press; 2007.

[2] Tichopad A, Kitchen R, Riedmaier I, Becker C, Stahlberg A, Kubista M. Design and optimization of reverse-transcription quantitative PCR experiments. Clin Chem 2009;55:1816–23.

[3] Kubista M, Andrade JM, Bengtsson M, Forootan A, Jonák J, Lind K, et al. The real-time polymerase chain reaction. Mol Aspects Med 2006;27(2–3):95–125.

[4] Mackay IM, Mackay F, Nissen MD, Sloots TP. Real-time PCR: history and fluorescent chemistries. In: Mackay IM, editor. Real-time PCR in microbiology from diagnosis to characterization. 1st ed. Poole, UK: Horizon Press-Caister Academic Press; 2007. p. 1–40.

[5] Livak KJ, Schmittgen TD. Analysis of relative gene expression data using real-time quantitative PCR and the 2(−delta delta C(T)) method. Methods 2001;25(4):402−8.

[6] Zuker M. Mfold web server for nucleic acid folding and hybridization prediction. Nucleic Acids Res 2003;31(13):3406−15.

[7] Bustin SA. A−Z of quantitative PCR. La Jolla, California: IUL Press.

[8] Bartlett JMS, Speirs H. Extraction of nucleic acid templates. Methods Mol Med 2004;97:59−70.

[9] Liu X, Harada S. RNA isolation from mammalian samples. Curr Protoc Mol Biol 2013; Chap 4:Unit 4.16.

[10] Riedmaier I, Bergmaier M, Pfaffl MW. Comparison of two available platforms for determination of RNA quality. Biotechnol Biotechnol Eq 2010;24(4):2154−9. Available here: http://rna-integrity. gene-quantification.info/.

[11] Bustin SA, Benes V, Garson JA, Hellemans J, Huggett J, Kubista M, et al. The MIQE guidelines: minimum information for publication of quantitative real-time PCR experiments. Clin Chem 2009;55(4):611−22.

[12] Vandesompele J, De Preter K, Pattyn F, Poppe B, Van Roy N, De Paepe A. Accurate normalization of real-time quantitative RT-PCR data by geometric averaging of multiple internal control genes. Genome Biol 2002;3:7.

[13] Andersen CL, Jensen JL, Orntoft TF. Normalization of realtime quantitative reverse transcription-PCR data: a model-based variance estimation approach to identify genes suited for normalization, applied to bladder and colon cancer data sets. Cancer Res 2004;64:5245−50.

[14] Schmittgen TD, Livak KJ. Analyzing real-time PCR data by the comparative C(T) method. Nat Protoc 2008;3(6):1101−8.

WEB RESOURCES

List and information about qPCR instruments: http://www.cyclers.gene-quantification.info/

Overview of available detection chemistries and considerations for chemistry choice: http://www.gene-quantification.de/chemistry.html

MIQE guideline checklist: http://medgen.ugent.be/rdml/miqe.php

CHAPTER 5

Metal Metabolism and Detoxification

Contributed by Joëlle Auclair and François Gagné

Chapter Outline

In this chapter, methods to determine the levels of labile metals, intracellular metal partitioning, and the levels of metallothioneins (MT) will be presented. MTs are low molecular weight (6 kDa), heavy metal binding proteins due to the high proportion of cysteine (20−30% of total amino acids). The lack of hydrophobic region because of the absence of nonpolar amino acids renders the protein heat stable with the absence of the usual absorbance at 280 nm for most proteins. In natural conditions, MT binds essential elements such as copper (Cu) and zinc (Zn). The presence of toxic metals such as mercury (Hg), cadmium (Cd), and silver (Ag) strongly binds to MT and liberates Zn and Cu, which, in turn, induce *de novo* synthesis of MT [1]. MTs also have the capacity to sequester reactive oxygen species such as nitric oxide and hydroxyl radicals [2] and the protein is also induced by oxidative stress [3]. During that process, a metal ion is exchanged with the reactive oxygen species from the protein complex where the resulting toxicity will depend on the nature of the released metal. For example, the release of Cu or Hg is much more damaging than the release of Zn from MT. In this context, the mobilization of metals is coupled to oxidative stress (see Figure 5.1).

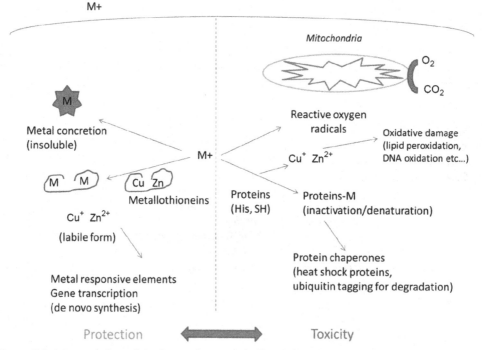

Figure 5.1 A general view of the fate and potential effects of metals in cells.

The ionic form of metals could bind to metal binding sites of proteins and release essential elements such as Zn^{2+} and Cu^+, which could either lead to cell damage or activate metal binding proteins such as MTs. A balance between metal protection and toxicity exists, which depends on the type and concentration of metals involved. For example, Cu and Cd lead to important oxidative stress and damage in cells and also bind strongly to MTs.

5.1 INTRACELLULAR FREE METAL AVAILABILITY

5.1.1 Introduction

The entry of nonessential metals in cells competes with the binding sites of essential metals, such as Zn and Cu. The release of ionic Zn represents an early biomarker of exposure to divalent metals in cells [4]. Labile Zn levels were significantly increased in rainbow trout hepatocytes exposed to Cd^{2+}, Hg^{2+}, Ag^{1+}, and Zn^{2+}. This assay is simple and rapid if the analyst has access to a fluorescence microplate reader. A cuvette-based fluorometer could also be used, although less quick to acquire the data. The assay is based on the Zn-induced fluorescence upon binding to a probe.

5.1.2 Reagents

Stock probe solution: Dissolve 25 mg of 6-methoxy-8-p-toluenesulfonamido-quinoline (TSQ) in 3 mL of dimethylsulfoxide (DMSO) 99.9%. Conserve 100 µL aliquots at 4°C in dark vials.

Reagent probe: Dilute TSQ stock solution to obtain 50 µM in 20% DMSO in phosphate buffered saline (PBS; 140 mM NaCl, 5 mM KH_2PO_4, pH 7.4). For example, 40 µL of stock probe is added to 20 mL of 20% DMSO in PBS. Prepare daily just before the assay.

Zn chloride solution: Dissolve 20.9 mg of $ZnCl_2$ in 10 mL of SQ water to give 1 mg/mL Zn. Prepare a 0.2 µg Zn/mL standard solution: 20 µL of 1 mg/mL in 100 mL of SQ water. Prepare daily.

5.1.3 Procedure

The assay is usually performed on cell suspensions, the cytosolic fraction of tissue homogenates (the supernatant after $100,000 \times g$ for 60 min at 4°C) or the S15 fraction (the supernatant after $15,000 \times g$ for 30 min at 4°C) at least.

In duplicate wells, 40 µL of SQ water is added to 10 µL of blank and S15. In another set of duplicate wells, 10 µL of standard $ZnCl_2$ solution is added to 10 µL of PBS (or S15 if standard addition method is recommended) and completed to 50 μL with SQ water. A volume of 150 μL of reagent probe is added to the wells (four wells per sample) and mixed for 10 min at room temperature. Fluorescence is read at 360 nm excitation (20–40 nm bandwidth) and 460 nm (20–40 nm bandwidth). The sensitivity setting of the instrument is programmed to read fluorescence values between 5 and 20 ng/mL Zn.

5.1.4 Data Calculation

$$\text{Zn concentration in sample (ng/mL)} = [\text{RFU sample} - \text{blank}/\text{RFU(spiked sample)}$$
$$- \text{blank}] \times \text{Concentration added standard (ng/mL)} \times \text{dilution factor,}$$

where RFU is the Relative Fluorescence Unit or the emission readings acquired by the instrument. The concentration of the added standard was 10 ng/mL in the present example and the dilution factor was $200/10 = 20$. The concentration of Zn is then normalized against total protein content or tissue weight to give ng Zn/mg proteins or ng Zn/g tissue. The Zn concentration could be obtained by the internal addition method if the sample matrix interferes with the fluorescence signal (quenching of excitation or emission photons), which commonly occurs during fluorescence measurements. The TSQ fluorescent signal could also be blocked by the addition of 0.2 mM of a nonfluorescent metal chelator such as N,N,N',N'-tetrakis(2-pyridylmethyl (TPEN) after making readings to determine the background autofluorescence in complex samples.

5.2 SUBCELLULAR METAL BINDING FRACTIONS

Another mechanism of metal inactivation or detoxification consists in the mobilization of metals in insoluble calcium concretions in tissues of invertebrates [5]. The sequestration of metals in calcium concretions in tissues was shown to reflect ambient metal exposure and metals trapped in these crystalline concretions are considered not biologically available. The decrease in metal concretions with the increase in the heat-instable protein (most proteins not containing MT) is considered a toxic response. This response might be tissue specific since not all tissues contain insoluble concretions. They are usually found at high quantities in gill tissues and somewhat in the digestive gland of bivalves. The assay is relatively simple and depends on the laboratory's capacity to analyze a suite of metals such as Cd, Cu, Zn, or other metal of interest. The metals could be conveniently determined by flame or graphite furnace atomic absorption spectrometry.

5.2.1 Reagents and Equipment

Homogenization buffer: Prepare 50 mM Tris-acetate at pH 7.2 (1 L). Dissolve 6 g of Tris base (2-amino-2-hydroxymethyl-propane-1,3-diol) in 900 mL of SQ water, adjust pH to 7.2 with 10% acetic acid, and complete to 1 L with SQ water.

Refrigerated microcentrifuge

Atomic absorption spectrometer

5.2.2 Procedure

The tissues are homogenized in ice-cold Tris-acetate buffer using a Teflon pestle (five passes) at a ratio of 20 g/100 mL buffer. The homogenate is decanted or passed through a sieve (100−200 μm) to remove tissue fragments. All steps should be performed at 4°C. The homogenate is centrifuged at $800 \times g$ for 15 min where the pellet and supernatant are separated and kept aside on ice. The metal concretions or the insoluble metal fraction is contained in the pellet while the heat-unstable metal fraction (protein) is contained in the supernatant as described below.

Metal concretion: The pellet containing the metal concretions should be washed by resuspending in an equal volume of homogenization buffer and recentrifuging at the same speed. The pellet is then heat treated at 100°C for 2 min, incubated in 1 M NaOH at 60−70°C for 10 min, and centrifuged at 10,000 g for 10 min. The pellet is kept aside for nitric acid digestion (2 M overnight) for metal analysis by atomic absorption spectrometry.

Heat-unstable metal fraction: The supernatant is centrifuged at $15,000 \times g$ for 30 min and the supernatant filtered on 0.2 μm cellulose acetate filter to remove remaining membrane vesicles or centrifuged at $100,000 \times g$ for 60 min. The filtered sample is heat treated at 80−90°C for 10 min, placed on ice for 10 min, and centrifuged at $10,000 \times g$ for 10 min at 2−4°C. The pellet is dissolved in 2 M nitric acid for 12 h to allow digestion and metals analyzed by atomic absorption spectrometry. MTs are considered the major *heat-stable* proteins, thus they remain in the supernatant; however, the supernatant also contains heat-stable peptides (glutathione) or amino acids (cysteine, histidine) capable of binding various metals and free metals. This fraction could be considered to harbor the dissolved fraction of metals in tissues if MT could be removed by ultrafiltration on 3-10 kDa membrane. The presence of metals in the supernatant could also be determined.

5.2.3 Data Analysis and Calculation

The amount of metals determined by atomic absorption spectrometry is corrected by the dilution factor: resuspension of pellet/volume for analysis (atomic absorption spectrometry). Metal analyses are usually done with standards of corresponding metal in 0.1–0.5 M nitric acid and the use of appropriate matrix modifiers as recommended by the supplier's instrument. If Zeeman-based or deuterium-based background corrections are used, the use of external standard curves is usually appropriate for diluted samples. If low dilutions are used and matrix interferences are suspected, then use internal standards or the standard addition method of calibration is recommended. Essentially, the absorbance is measured before and after the addition of a standard concentration directly in the sample to take into account matrix interferences. The concentration is then normalized by the tissue weight to give μg metals/g tissue weight.

5.3 METALLOTHIONEINS

5.3.1 Silver Saturation Assay

The silver saturation assay represents a robust, specific, and sensitive way to determine the levels of MT in organisms. Although this assay is generic, i.e., it is based on the special properties of the protein such as thermal stability and the very strong binding affinity for monovalent Ag at alkaline pH, it is considered the most specific and sensitive of the indirect assays for MT. The silver saturation assay was introduced as a sensitive and specific means for MT evaluations since 1 mol of Ag^+ binds 1 mol of cysteine and there are 15–17 cysteine residues per MT molecule. Ag^+ binds thiols (R-SH) very strongly at pH 8.5 and is able to displace other metals that bind strongly to thiols such as Cu and Hg. Ag is more strongly retained by MT during hemoglobin-scavenging steps. Hence, the method is less prone to interference by heat-stable, low molecular weight thiols (glutathione or cysteine) since the addition of excess hemoglobin strips loosely bound Ag. The original methodology used radioactive Ag [6], which was followed by a nonradioactive one based on graphite furnace atomic absorption spectrometry [7]. For a review on the numerous methods for MT assessments, the reader is invited to consult recent reviews [1,8].

5.3.1.1 Reagents

Glycine buffer: Dissolve 1.5 g glycine (sodium salt) in 190 mL of water (0.1 M concentration), adjust the pH to 8.5 with NaOH 1 M, and complete to 200 mL with SQ water.

AgNO₃ reagent: A 1000 ppm stock solution of Ag^+ is prepared by adding 0.157 g $AgNO_3$ in 100 mL of 0.1 N of HNO_3. Concentrated solutions of Ag are usually commercially available at usual chemical suppliers. The stock solution is serially diluted in glycine buffer to yield 10 ppm (prepare daily).

Hemoglobin solution: Dissolved 0.2 g of bovine hemoglobin in 10 mL of water (prepare daily), keep at 4°C.

Matrix modifier: Dissolve 0.1 g of diammonium phosphate $((NH_4)H_2PO_4)$ in 10 mL bidistilled water.

5.3.1.2 Procedure

The tissue homogenate is prepared in 0.2 M sucrose containing 10 mM Tris-acetate, pH 7.5, 5−10 mM β-mercaptoethanol or dithiothreitol (DTT), and 0.1 mM phenylmethylsulfonyl fluoride (PMSF) or 1 μg/mL protease inhibitor (apoprotinin). The homogenate is centrifuged at $15,000 \times g$ for 20 min at 2°C and the supernatant (S15) carefully collected from the upper lipid layer and pellet.

One volume of $AgNO_3$ reagent is added to S15 (10−50 μL) in a screw cap microcentrifuge tube for 10 min at room temperature (silver saturation step). The volume is adjusted to 250 μL with glycine buffer and 50 μL of hemoglobin solution is added and allowed to incubate for 15 min at room temperature (excess silver removal step). The mixture is then placed in a boiling water bath for 2 min, cooled at room temperature for 5 min, and centrifuged at $10,000 \times g$ for 5 min. The hemoglobin addition step is repeated once more to the supernatant and the supernatant collected for total Ag analysis.

Total Ag could be determined by a variety of methodologies but graphite furnace atomic absorption spectrometry is the method of choice because of the low sample volume needed to perform the analysis (10−30 μL) and low detection limit (0.1−0.5 ng/mL) for this element. The sample is diluted 1/10 in water and the matrix modifier is 1% ammonium diphosphate monobasic $((NH_4)H_2PO_4)$. The resonance line for Ag is 328.8 nm and the atomization temperature is set at 1900°C with an ashing temperature of 600°C for this matrix. A calibration curve for Ag is also constructed with the instrument between 1−10 ng/mL Ag.

5.3.1.3 Data Analysis

The amount of Ag in ng/mL is multiplied by the dilution factor (250 μL/sample volume μL) and by the dilution factor for Ag analysis by atomic absorption spectrometry ($\times 10$). The concentration of Ag is then converted in nanomoles by dividing with the atomic weight of Ag (107.86 g/mol).

Because 17 Ag atoms bind 1 MT macromolecule, the Ag concentration (nmol/mL) is divided by 17 to give nmol MT/mL. This ratio was confirmed by rabbit and rat liver MT-II. The ratio number could be changed/corrected by the amount of cysteine residues in MT if known for the particular species (PubMed, protein sequences).

The concentration of MT in nmol/mL is normalized by the protein content in the S15 fraction or by the tissue weight. The data are expressed as nmol MT/mg proteins or nmol MT/g tissue weight.

5.3.2 Spectrophotometric Assay

The determination of thiol contents in MT could be conveniently determined by spectrophotometry [9], which makes this assay very accessible to many laboratories regardless of the test species, albeit much less sensitive than the silver saturation assay. In this modified approach, the MT proteins are first reduced by a strong reducer before the selective solvent fractionation and precipitation steps with organic solvents, because MT is highly prone to oxidation during isolation. The isolated MT fraction is then treated to low pH in the presence of chelator and high salt to displace metals and maintain the apo-MT for the spectrophotometric determination of thiols (R-SH) using Ellman's reagent.

5.3.2.1 Reagents

Tris-acetate-NaCl buffer: Dissolve 12.11 g Tris base (100 mM) and 110.88 g NaCl (2 M) in 900 mL SQ water. Adjust pH to 8.2 with acetic acid 20% and complete to 1 L with SQ water.

Ellman's stock solution: A stock solution of 5 mM 5,5'-dithiobis-(2-nitrobenzoic acid) (DTNB) is prepared in 200 mM Tris-acetate, pH 8.2, containing 0.1 mM EDTA. A 100 mL stock is prepared as follows: 2.422 g Tris base and 4.5 mg EDTA are dissolved in 80 mL SQ water and the pH adjusted to 8.2 with acetic acid 20%. Dissolve 0.2 g of DTNB and complete to 100 mL with SQ water. Store at 4°C in the dark for up to 2 months.

Ellman's reagent: Prepare 3.2 mL of Ellman's stock with 36.8 mL of Tris-acetate-NaCl buffer. Keep out of light.

Phosphine reagent: Dilute stock phosphine (Tris(2-carboxyethyl)phosphine hydrochloride) solution (0.5 M) 1/10 in Tris-acetate-NaCl buffer to obtain a final concentration of 50 mM. Keep from light.

GSH standard: 3 mg GSH in 10 mL HCl (0.1 M). Stable for one week, protect against light.

Ethanol-8% chloroform: Mix 46 mL of 95% ethanol with 4 mL of chloroform in a fume hood.

Ethanol 87% and chloroform 1%: Mix 43.5 mL of 95% ethanol with 0.5 mL chloroform and 6 mL of SQ water.

NaCl-EDTA solution: Dissolve 1.46 g of NaCl and 0.18 g EDTA in 100 mL water.

RNA solution: Dissolve 50 mg of total RNA in 50 mL of NaCl-EDTA prediluted 1/5 with SQ water. The sample could be heated at 60–70°C to assist dissolution.

HCl solution: Pipette in a fume hood with gloves 5 mL of concentrated HCl in 10 mL of SQ water (6 N final concentration).

5.3.2.2 Procedure

Tissues are homogenized with a Teflon pestle tissue grinder apparatus in 2–3 volumes of 140 mM NaCl saline buffered with 10 mM Tris-acetate, pH 7.5, containing 0.5–5 mM DTT or 1–10 mM β-mercaptoethanol and 1 µg/mL apoprotein or 1 mM PMSF as protease inhibitor. The homogenate is then centrifuged at 15,000 × g for 30 min at 2–4°C and the supernatant (S15) collected.

The S15 sample (200 µL) is pretreated with 50 µL of 50 mM phosphine for 15 min to maintain reductive condition in the S15 sample. One volume of ethanol-chloroform (92:8) is added, mixed for 1 min, and centrifuged at 6000 × g for 10 min at 4°C. A volume of 400 µL of the supernatant is collected and mixed with 50 µL of RNA and 10 µL of HCl. The addition of HCl drops the pH to release metals from MT (apo-MT). RNA is added as a carrier to coprecipitate with MT. Ethanol (1 mL) containing 1% chloroform is added to the mixture and centrifuged again at 6000 × g for 10 min at 4°C. The supernatant is discarded and the pellet washed in ethanol. The pellet is resuspended in 150 µL of 0.25 M NaCl and 4 mM EDTA and transferred to clear microplates. A 70 µL volume of the resuspended pellet is mixed with 30 µL of 0.1 M HCl and allowed to stand for 5 min before adding 30 µL Tris-acetate buffer for another 5 min. A volume of 200 µL of Ellman's reagent is then added in each microplate. Absorbance is measured against a standard concentration of reduced glutathione (10 nmol/mL or 20 nmol/mL) at 412 nm. The blank contains only 130 µL of solution composed of 0.25 M NaCl-4 mM EDTA, HCl, and Tris-acetate buffer as described above.

5.3.2.3 Data Calculation, Normalization, and Comments

Reduced thiol content nmol/mL = [Asample − Ablank]
× [GSH concentration (nmol/mL)/Astandard − blank] × dilution factor

The dilution factor is calculated as follows: $(500/200\,\mu L\ S15) \times (400\,\mu L/150\,\mu L) \times (330\,\mu L/70\,\mu L) = 31.4$.

The relative amount of MT could be calculated if the number of cysteine molecules is known for the given test species. The amino sequence for MTs could be found through PubMed (http://www.ncbi.nlm.nih.gov/pubmed?cmd = search&db = PubMed). For example, rat MT contains 19 cysteines per MT: MT nmol/mL = reduced thiol content/19. The data are usually expressed as nmol thiol/mL when this ratio is unknown for the given test species. However, in the original methodology about 70% of the thiol residues reacted with Ellman's reagent while the remaining 30% thiols were not detected by this methodology. Hence, the MT values could be underestimated by a factor of 30% with this assay.

The amount of reduced glutathione equivalents is then normalized against total proteins in the S15 (mg/mL) or tissue weight (g/mL) to give nmol thiols/mg proteins or nmol thiols/g tissues.

5.4 ENZYME IMMUNOASSAY

MT antibodies are readily available by many commercial suppliers. These assays are expected to be more specific and highly sensitive if background antibody nonspecific binding is well controlled and the chemiluminescent or fluorescent substrates are used. In this chapter, a direct immunoassay for fish MT is proposed using a very sensitive chemiluminescent detection system in microplates.

5.4.1 Reagent

PBS: Prepare a 1 L solution of 140 mM NaCl and 5 mM KH_2PO_4 and adjust to pH 7.4 with NaOH 1 M.
PBS-Tween 20: Add 0.1 g of Tween 20 to 100 mL of PBS (PBS-Tween at 0.1%).
Blocking solution: Dissolve 3 g of serum bovine albumin or fat-free milk powder, 0.01 g of Tween 20, and 1.5 mg DTT to 100 mL of PBS. Dissolve

thoroughly with a magnetic stirrer and filter on 0.45 or 0.2 μm pore filter. Stable for 3 days when stored at 4°C in the dark.

Sodium bicarbonate: A 50 mM solution of $NaHCO_3$ is prepared and the pH adjusted to 9.6.

Primary antibody: Dilute commercial mouse anti IgG against trout MT 1/500 in PBS–Tween 20. Secondary antibody with peroxidase for mouse IgG antibodies. Dilute commercial preparation 1/1000 in PBS–Tween 20.

Peroxidase reagent: Prepare fresh 10 μM luminol and 100 μM H_2O_2. Mix together, keep on ice in the dark.

5.4.2 Procedure

The assay is performed on the S15 (see above) fraction of liver or gill homogenates of rainbow trout. The S15 fraction is diluted to obtain 20 μg/mL with sodium bicarbonate buffer, 50 μL (1 μg total protein) is added in special high binding immunoplates (Immunolon 96-well microplates, 8-well inserts), and the wells are covered and stored at 4°C overnight to allow binding of proteins to the wells. A setting time of 3 h is enough for protein absorption to the wells at room temperature. After this incubation period, the solution is removed by aspiration, washed with 150 μL PBS, and 150 μL of blocking solution is added to wells. After incubation at room temperature for 30–60 min, the blocking solution is removed by aspiration and washed twice with PBS. The primary antibody is added to each well (100 μL) and incubated for 60 min at 37°C. After this incubation period, the wells are washed twice with 150 μL of PBS and 100 μL of secondary antibody is added. After incubating for 30 min at 37°C, the cells are washed three times with PBS. The peroxidase reagent solution is added in each well (5–100 μL) and luminescence read at 5 min intervals for 20 min. Blank solution on the wells consists of sodium bicarbonate buffer only.

The assay could also be practiced in cells instead of S15 fraction. The sodium bicarbonate fixation step is replaced by cell fixation step where 20,000 cells are fixed in PBS containing 5 mM $MgCl_2$ and 2% formaldehyde for 5–10 min in normal white microplates (centrifuged at $200 \times g$ for 5 min). The fixative is removed and 150 μL of PBS containing 0.5% albumin and 0.2% Tween 20 is added for 30 min to permeabilize the cells, block nonspecific binding sites, and quench the trace of fixative. The antibody solutions are prepared in PBS containing 0.5% albumin and 0.1% Tween 20 is used for the following steps as described above.

5.4.3 Data Calculation and Analysis

The increase in light emission over time is proportional to the amount of peroxidase enzyme. Peroxidase activity $= (\mathrm{RL}t2 - \mathrm{RL}t1)/(t2 - t1)$ where RL is the relative light emission. Time $t1$ corresponds to the initial reading at $0-1$ min and $t2$ at 10 or 15 min. The linear portion of the increase in light emission should be selected.

5.5 HEAT SHOCK PROTEIN 70 DETECTION AND QUANTIFICATION BY WESTERN BLOT ANALYSIS

Contributed by Joelle Auclair

The heat shock protein 70 family (Hsp70) is composed of highly conserved proteins that are expressed ubiquitously in all organisms [10]. Some of their members, such as hsp72, are induced upon a stress event (i.e., heavy metals, extreme temperature, oxidative stress, etc.) and act as cytoprotective agents by targeting misfolded proteins [11−13]. Since heat shock proteins play a major role in cellular homeostasis based on their chaperone capacity, the increased expression hsp70 proteins represent a key biomarker to assess the impact of environmental chemical insults [14,15].

Different techniques could be used to quantify protein expression levels. Among them, Western blot, also called protein immunoblot, has been widely used for protein analysis in different research and diagnostic fields. This technique is based on the electrotransfer of proteins from polyacrylamide gel to a membrane support, which is followed by antibody recognition allowing specific detection of a protein of interest [16]. Western blot relies on a series of steps that begins with sample preparation and ends with signal detection and semiquantitative analysis. A detailed overview of the Western blot procedure is outlined in Figure 5.2.

This section proposes a Western blot protocol for Hsp72 quantification suitable for freshwater bivalves, which can be easily adapted and applied to other species. The protocol also integrates explanatory notes to provide additional support and tables with useful technical information for specific optimization and troubleshooting.

5.5.1 Reagents and Materials

Homogenization buffer: Prepare 1 L of 100 mM NaCl, 25 mM HEPES-NaOH, pH 7.4, 1 mM DTT, and 1 μg/mL aprotinin by dissolving 5.488 g

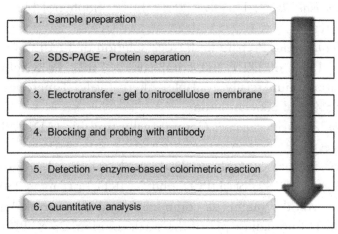

1. Sample preparation

2. SDS-PAGE - Protein separation

3. Electrotransfer - gel to nitrocellulose membrane

4. Blocking and probing with antibody

5. Detection - enzyme-based colorimetric reaction

6. Quantitative analysis

Figure 5.2 Western blot analysis. Procedure overview.

NaCl, 5.9578 g HEPES, and 0.0154 g DTT. Adjust pH to 7.4 with NaOH 1 M and complete to 1 L with distilled water. Homogenization buffer is stable for 6 months when stored at 4°C. Add fresh aprotinin (protease inhibitor) on the day of homogenization.

Protein quantification reagent: Bradford assay dye reagent (commercially available) or alternatively, reagents for Lowry, bicinchoninic acid (BCA), or Biuret total protein quantification assays.

Sodium dodecyl sulfate (SDS) or lithium dodecyl sulfate (LDS)—Sample loading buffer: Ready to use 4× and 2× concentrates are available commercially. Alternatively, classical 2× Laemmli sample buffer (125 mM Tris-HCl, pH 6.8, 4% SDS, glycerol 20%, and 0.004% bromophenol blue) can be used for standard SDS-PAGE. Prepare 2× Laemmli sample buffer by mixing 1.25 mL 1 M Tris-HCl, pH 6.8, 4 mL 10% SDS, 2 mL glycerol, and 0.2 mL bromophenol blue 0.2%. Complete to 9 mL with distilled water.

Add fresh the reducing agent: 100 µL of 1 M DTT for each 900 µL of 2× Laemmli sample buffer or 100 µL of β-mercaptoethano1.

1 M Tris-HCl, pH 6.8: Dissolve 6 g Tris base in 30 mL of distilled water. Adjust pH to 6.8 with 1 M HCl. Complete to 50 mL with distilled water. Store at 4°C.

10% SDS (stock solution): Dissolve 1 g of SDS in 8 mL of distilled water then complete to 10 mL with distilled water. Stable for at least 6 months when stored at room temperature.

Bromophenol blue 0.2%: Dissolve 20 mg of bromophenol blue in 10 mL of distilled water.

Sample reducing agent: Prepare fresh stock solution of 1 M DTT. Add distilled water to 0.154 g of DTT and complete to 1 mL. Preweight formats of DTT are also commercially available. Alternatively, β-mercaptoethanol can be used at 5% final concentration.

Hsp70/Hsp72 protein standard: Commercially available (ADI-SPP-763; Enzo Life Sciences).

25× PBS: Prepare 500 mL of 25× stock solution by dissolving 100 g NaCl, 2.5 g KCl, 18 g Na_2HPO_4, and 3 g KH_2PO_4 in ~250 mL of distilled water. Mix thoroughly all components and fill up to 500 mL with distilled water. Do not adjust pH of the stock solution. 25× PBS is stable for at least 6 months when stored at room temperature.

PBS 1×: Prepare 1 L of fresh PBS 1× working solution (137 mM NaCl, 2.7 mM KCl, 10 mM Na_2HPO_4, and 1.8 mM KH_2PO_4, pH 7.4) by diluting 40 mL of 25× PBS in 960 mL of distilled water. The pH value should be near 7.4. Adjust pH value if necessary. Keep at room temperature for the day of the experiment. Remaining solution can be stored a few days at 4°C.

10× Tris-buffered saline (TBS; 10×): Prepare 1 L of 10× stock solution by dissolving 24.2 g of Trizma base and 87.6 g of NaCl in 800 mL of distilled water. Adjust pH to 7.5 with HCl and complete to 1 L with distilled water. 10× TBS is stable for at least 6 months when stored at 4°C.

TBS 1×: Prepare 500 mL of fresh TBS 1× working solution (20 mM Tris and 150 mM NaCl, pH 7.5) by diluting 50 mL of 10× TBS in 450 mL of distilled water. The pH value should be near 7.5. Adjust pH value if necessary. Keep at room temperature for the day of the experiment. Remaining solution can be stored a few days at 4°C.

Prestained protein molecular weight marker (10−250 kDa range): Commercially available (e.g., Precision Plus Protein Kaleidoscope Standard from Bio-Rad or equivalent).

SDS-polyacrylamide gel: Precast 4−12% Bis−Tris mini gel (commercially available). Alternatively, equivalent precast mini gels or 10% acrylamide gels can be used. For handcasted gels, Laemmli protocol for Tris−glycine SDS-polyacrylamide gels can be used [8].

SDS running buffer: If using commercial precast gels, follow the manufacturer's recommendations for an optimal choice of running buffer. Otherwise, for electrophoresis that is performed with Tris-glycine SDS-polyacrylamide gels, Tris-glycine-SDS running buffer should be used.

Laemmli running buffer (Tris 25 mM, glycine 192 mM, and SDS 0.1%): Prepare Tris-glycine-SDS running buffer by dissolving 1 g SDS, 3.03 g Tris base, and 14 g glycine in 900 mL of distilled water. Mix and complete to 1 L with distilled water. The final pH should be 8.3. Do not adjust pH. Discard buffer if a different pH value is observed.

Transfer buffer: For a system using commercial precast gels, follow the manufacturer's recommendations for an optimal choice of transfer buffer. Otherwise, for Laemmli gels, Tris-glycine buffer (25 mM, 192 mM, 20% methanol) should be used. For a wet transfer, volume to be prepared depends on the type of transfer tank, but in most of the cases 2 L of transfer buffer is required. Prepare fresh Tris-glycine buffer by dissolving 6.06 g Tris base and 28.8 g glycine in 1 L of distilled water, then add 400 mL of methanol (analytical grade). Complete to 2 L with distilled water. The final pH should be near 8.3. Do not adjust pH. Discard buffer if a different pH value is observed. Prechill the buffer a few hours before transfer by placing it at 4°C.

Nitrocellulose membrane (0.2 μm pore size): Commercially available in different packaged precut sheets and rolls. When cutting nitrocellulose membranes, make sure that they have the same dimensions as the gel.

Blotting filter paper: Commercially available in different sizes and ready-to-use precut formats. If blotting filter papers need to be cut, make sure that they have the same dimensions as the gel.

Ponceau S solution: Ready-to-use formats are commercially available. Prepare 0.1% Ponceau S in 5% acetic acid by dissolving 0.1 g of Ponceau S in 100 mL of 5% acetic acid aqueous solution. Keep this ready-to-use solution at room temperature.

Blocking solutions: Prepare 100 mL of fresh blocking solution for both target proteins (Table 5.1).

Primary antibody solution: Prepare 5 mL of fresh primary solution for both target proteins as explained in Table 5.2.

Secondary antibody solution: Prepare 5 mL of secondary solution for both target protein as shown in Table 5.3.

Table 5.1 Preparation of Blocking Solutions for Western Blot

Protein	Blocking Agent	Preparation
Hsp70	Dry milk 1% in PBS 1 ×	Dissolve 1 g of dry milk in 80 mL of PBS 1 ×. Complete to 100 mL with PBS 1 × and mix thoroughly. Keep remaining solution on ice for the day of experiment.
Actin	Dry milk 5%/Tween 0.5% in PBS 1 ×	Dissolve 5 g of dry milk in 80 mL of PBS 1 × and add 500 μL of Tween 20. Care must be taken to avoid formation of bubbles. Complete to 100 mL and mix gently. Keep remaining solution on ice for the day of experiment.

Table 5.2 Preparation of Primary Antibody Solution at the Day of Analysis

Protein	Primary Antibody	Dilution	Preparation
Hsp70	Rabbit anti-hsp-70 (SPA-812) polyclonal antibody (Enzo Life Sciences)	1:1000	Add 5 μL of anti-hsp70 antibody to 5 mL of Hsp70 blocking agent (dry milk 1% in PBS 1 ×) and mix. Use immediately.
Actin	Mouse anti-actin (clone Ac-40) monoclonal antibody (Sigma)	1:1000	Add 5 μL of anti-actin antibody to 5 mL of actin blocking agent (dry milk 5%/Tween 0.5% in PBS 1 ×) and mix. Use immediately.

Table 5.3 Preparation of Secondary Antibody Solution at the Day of Analysis

Protein	Secondary Antibody	Dilution	Preparation
Hsp70	Goat anti-rabbit IgG, alkaline phosphatase conjugate	1:1000	Add 5 μL of anti-rabbit antibody to 5 mL of Hsp70 blocking agent (dry milk 1% in PBS 1 ×) and mix. Use immediately.
Actin	Goat anti-mouse IgG, alkaline phosphatase conjugate	1:1000	Add 5 μL of anti-mouse antibody to 5 mL of Hsp70 blocking agent (dry milk 1% in PBS 1 ×) and mix. Use immediately.

Colorimetric alkaline phosphatase substrate (NBT/BCIP): Commercially available (e.g., AP Conjugate Substrate Kit from Bio-Rad or equivalent). Follow manufacturer's recommendations for substrate preparation and use.

Special equipment:

Water bath or Thermomixer

Platform shaker

Polyethylene bags

Heat sealer for polyethylene bags

Vertical electrophoresis chamber (e.g., mini gel systems, XCell SureLock Mini-Cell from Invitrogen)

Transfer unit (e.g., mini gel systems, XCell II Blot Module, Invitrogen)

Power supply

Polytron homogenizer

Flatbed scanner

Image analysis software.

5.5.2 Procedure

5.5.2.1 Sample Preparation

Homogenize tissues in cold buffer (100 mM NaCl, 25 mM HEPES-NaOH, pH 7.4, 1 mM DTT, and 1 µg/mL aprotinin) in a 1:5 ratio (1 g tissue/5 mL buffer) with a Polytron homogenizer. The use of proper extraction buffer with protease inhibitors will ensure, while cell disruption will occur, that pH is maintained and endogenous proteolytic enzymes are neutralized. Mechanical disruption will tend to generate some heat and working on ice is a prerequisite. Centrifuge homogenates at $15,000 \times g$, 20 min at 4°C. Collect supernatants, which contain soluble proteins, and discard pellets. If necessary, supernatants can be stored at -80°C for later analysis. Hsp70 proteins are mainly found in cytosol compartment. Determine total protein content using the method of Bradford as explained elsewhere.

5.5.2.2 Sample Preparation for Polyacrylamide Gel Electrophoresis

It is important that equal amounts of total protein are loaded to each well. This will ensure good reproducibility of the procedure and the band densities will be directly related to protein expression and not to variation in added protein. For Hsp70, good signal detection can be obtained within a loading range of 5 to 20 µg total protein per well.

First, condition the samples for gel electrophoresis by mixing them with the loading buffer and reducing agent. Here, the amount of added samples needs to be adjusted to obtain the same final protein concentration (usually between 5 and 20 µg per lane). If using the $4 \times$ loading buffer, refer to Table 5.4 for additional information regarding volume of reagents required for sample preparation.

Prepare Hsp70 protein standard and load 2 µg per well. Proceed as explained previously for sample preparation. Once all the components are

Table 5.4 Sample Preparation, Electrophoresis Under Reducing and Denaturing Conditions

	Volume
Sample	$\times\ \mu L$
4 × loading buffer	5 μL
10 × reducing agent (DTT 500 mM)	2 μL
Distilled water	up to 13 μL
Total volume	20 μL

Note: Loading buffer contains glycerol to allow samples to reach the bottom of the well, a dye to follow up migration progress, and a detergent (SDS or LDS) to denatured proteins. Since Western blot analysis is usually conducted under reducing condition, DTT or β-mercaptoethanol (reducing agents) are added to samples. By cleaving disulfide bonds, the reducing agent will affect the tertiary and quaternary structure of proteins and therefore help to linearize their polypeptides chains.

added, briefly vortex each sample and heat at 70°C for 10 min. Let the samples cool down at room temperature for 5 min before loading the gel. Do not place them on ice because SDS could precipitate at cold temperatures.

5.5.2.3 Running Polyacrylamide Gel Electrophoresis

Predicted molecular weight of the target protein needs to be taken into consideration for gel selection. Polyacrylamide gels can be considered as a filter with pore size that can be modulated by changing acrylamide percentage. Higher percentage will generate smaller pore size and will preferentially be used to separate protein with low molecular weight. Gels with gradient can be used for a broader range of protein size and they have higher resolution than constant acrylamide percentage.

Protein separation is performed under reducing and denaturing conditions by SDS-PAGE. This type of electrophoresis enables a differential migration based on linear chain length. The presence of SDS in the protein samples, polyacrylamide gels, and running buffer will add a net negative charge on the protein and abolish the 3D conformation of proteins and unravel the proteins.

Take a precast 4–12% Bis–Tris mini gel. Rinse the gel with distilled water to remove any preservative buffer. Take off the white adhesive. Pull the comb out of the gel carefully. Using a transfer pipette, rinse all wells three times with the running buffer. Place the gel in the electrophoresis chamber and with a buffer dam (one gel run) form an inner sealed compartment. Add running buffer in the central compartment. Check for any leaking. Make sure that all wells are filled and buffer level is high enough. Then add the remaining buffer in the outer compartment. Load samples on the gel (usually between 5 and 20 µg total protein per well), a molecular weight marker, and an internal control (Hsp70 protein standard). When loading a gel, in addition to samples, it is important to include a molecular weight ladder/marker for subsequent protein identification and if possible an internal control that will give a positive signal. This control can be, for example, a purified protein purchased commercially or a sample generated with cell or tissue that is known to have a high expression level of the protein of interest.

Place the lid on top and connect it to a power supply. If using mini-gel electrophoresis systems, run the gel for ~1 h at 180 V. Ensure that migration is progressing well by looking at the front dye (blue dye in samples). Sample should enter the gel and migrate down toward the positive end. Stop electrophoresis when samples have nearly reached end of the gel.

5.5.2.4 Electrotransfer of Proteins—from Gel to Nitrocellulose Membrane

Once the electrophoresis is completed, remove the gel from the cassette then cut off the wells (stacking gel) and bottom end (foot). Place it in a container filled with transfer buffer and allow to equilibrate for 10 min. Wear clean gloves and manipulate the gel carefully. Polyacrylamide gel can break easily.

During that time, in a Pyrex tray, add enough transfer buffer to soak the nitrocellulose membrane and sponge pads. Filter papers are soaked as well. Assemble all components (sponge pad, filter paper, membrane, and gel) in the transfer compartment to form a sandwich (Figure 5.3). First, start with sponge pads then add a filter paper and the gel. Complete assembling by adding sequentially a nitrocellulose membrane, filter paper, and sponge pads. Ensure

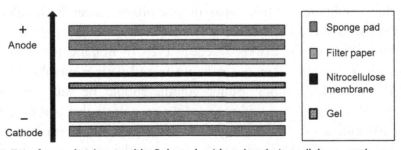

Figure 5.3 Transfer sandwich assembly. Polyacrylamide gel and nitrocellulose membrane.

that no air bubbles are trapped under layers. Air impedes current flow and consequently protein transfer. A glass culture tube can be used to remove them by delicately rolling on top of the sheets. It is very important to respect the stacking order of each layer and have a well-oriented sandwich in the transfer compartment. Negatively charged proteins contained in the gel will migrate toward the positive end of the compartment where the nitrocellulose membrane is placed.

Add transfer buffer as recommended by the manufacturer. Place the lid on top and connect it to a power supply. Proceed to the gel transfer for ~1 h at 30 V. If a current rise occurs, voltage can be lowered to prevent excessive heating. Once the transfer is completed, disassemble the sandwich and place the membrane in a tray containing distilled water. Rinse quickly two times to remove transfer buffer. Caution must be taken for membrane manipulation. Always wear clean gloves and use tweezers with a flat end.

The nitrocellulose membrane could be prestained with Ponceau S to determine the success of protein transfers to the nitrocellulose membrane [17]. This is achieved by adding enough staining to cover the membrane (work in a ventilated fume hood), and incubate for 1 min with low agitation. Pour off Ponceau S into a waste container, and rinse the membrane until the background color returns to white. As staining occurs, the protein bands should appear pink-red on a white background. If bubbles were present during the electrotransfer, they should form white spots on the membrane corresponding to an absence of protein. Electrophoresis/transfer problems are usually noticed at this step; refer to troubleshooting for more details.

5.5.2.5 Protein Revelation with Specific Antibodies

For the Hsp70 western blot, actin protein (42 kDa) is used as loading control. To achieve this double protein detection, the membrane needs to be cut. Protect the membranes with transparent cellophane, cover both sides. On a cutting mat, with the use of a scalpel and rule, do a straight cut just over the 50 kDa molecular weight marker. Identify each membrane for their specific probing (hsp70 or actin). Remove cellophane and destain membranes in PBS for 3×10 min soaking times. Buffer volume should be sufficient to cover membranes. Ponceau S is an acid reagent; the remaining dye will tend to react with the blocking agent by forming clots. From now on, never let the membranes dry out until the end of the immunodetection procedure. Place membranes in the proper blocking solution and incubate for 1 h under agitation at room temperature. Before the blocking step ends, prepare 5 mL of primary antibody solution. Place blocked membranes in plastic polyethylene bag and add 5 mL of the antibody dilution. Remove air bubbles by carefully pushing them through the bag opening. Properly seal the bag and incubate under agitation for 3 h at room temperature. After the incubation with the primary antibody, place membranes in containers (glass Petri dishes or plastic box covers) and rinse the membranes three times for 10 min with PBS. Several washes are necessary to remove unbounded antibodies. Omission or poor washing will induce unspecific background. Before the end of the washing step, prepare 5 mL of secondary antibody solution as described previously. Replace membranes in new plastic bags and add the secondary antibody. Remove any excess of air as indicated earlier and incubate for another 1 h at room temperature under light agitation. The secondary antibody needs to be specific to the species and the antibody class (i.e., I, G, E) from which the primary antibody has been raised (for example, for an antibody of class G developed in a rabbit use an anti-rabbit antibody for IgG). Rinse membranes 6×5 min with Tris buffer $1 \times$ under agitation (20 mL each wash).

5.5.3 Detection-Colorimetric Alkaline Phosphatase Activity

In Western blot, different methods can be used to reveal the resulting protein expression. Among the most popular, there are colorimetric (alkaline phosphatase conjugate antibody), chemiluminescent (horseradish peroxidase conjugated antibody), and fluorescent (fluorochrome (CyDye) labeled antibody) detection. The colorimetric method using NBT/BCIP was chosen for its sensitivity and because it does not require additional equipment for signal capture.

1. From the AP Conjugate Substrate Kit, prepare alkaline phosphatase buffer by adding 1 mL of alkaline phosphatase $25 \times$ buffer to 24mL of distilled water.
2. Prepare the substrate solution by adding 250 µL of reagents A and B to the alkaline phosphatase buffer (1:100 dilution).
3. Discard TBS $1 \times$ and add substrate solution to membranes.
4. Incubate under light agitation until bands appear clear and well defined. Be careful not to over incubate. Over incubation will cause a high background and lower signal-to-noise ratio.
5. Stop reaction by placing membranes in distilled water. Rinse 3×5 min in water.

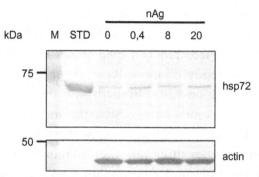

Figure 5.4 Expression of hsp70-inducible proteins (hsp72) in tissues of freshwater mussels *Elliptio complanata* exposed to increasing concentration of nanosilver (μg/L). Actin is used as a loading control. M, molecular weight maker; STD, Hsp70 protein standard.

The expression of hsp70-inducible proteins (hsp72) should appear below the 75 kDa molecular weight marker and have protein size comparable to the hsp70 protein standard (Figure 5.4). Actin proteins are detectable at ∼42 kDa.

Protein expression can be quantified by Western blot by using densitometry where each visible band can be transformed to a numerical value (pixel). The image of the scanned membrane is then treated by software able to detect difference in color density. The following steps to achieve a good quantification are provided.

Place the membranes in a clear transparent protector. On a flatbed scanner, scan them in grayscale at 400 dpi or with higher resolution (picture quality request). Save the files under the .tif image format. Proceed with densitometry using an image analysis program (ImageJ, an open access software, is commonly used). Always use grayscale image for quantification. If the original file is in color, it is possible to convert it into a black and white image with the software. Procedures for densitometry analysis are software specific. For subsequent steps and detailed protocol, refer to the software manual. Once all density values are generated for each band, it is important to normalize all hsp70 values with their corresponding actin loading control. This step will correct for signal variation attributed to loading inconstancy. As an alternative to actin, Ponceau S staining can also be used. For inter-gels/blots comparisons, each gel should be compared with a standard sample (e.g., Hsp70 or actin standard). This standard will enable us to account for the variation generated by the entire blotting

Table 5.5 Key Steps for Protocol Optimization

Parameters	Modifications
Blocking agent	Type (with or without detergent)
	Concentration (0.5–10% range)
	Incubation time (1 h to overnight)
	Temperature (usually, 4°C or room temperature)
Washing buffer	Type (TBS or PBS, with and without detergent)
	Incubation time (usually 5–10 min, can be longer)
	Number of washes (usually three or more)
Antibody dilution buffer	Type (with or without detergent)
	Concentration (0.5–10% range)
Primary antibody	Concentration (usually, 1:1000 or more; refer to manufacturer's recommendation)
	Incubation time (1 h to overnight)
	Temperature (usually, room temperature, and 4°C for overnight incubation)
Secondary antibody	Concentration (usually, 1:1000 or more; refer to manufacturer's recommendation)

and probing procedures (for example, blocking, washing, and antibody binding efficiency). For each blot, use the corresponding standard to normalize hsp70 relative density values. Finally, results are expressed as the relative amount of inducible hsp70 relative to actin.

For specific needs, as for working with a different species, it may be necessary to change a few parameters if results are not satisfactory. Listed in Table 5.5 are Western blot parameters that can be modified.

To test those modifications, it may be necessary to use a strip strategy that consists of loading the same sample throughout the gel and after transfer the membrane is cut into strips where each of them represents an entire well. This approach enables us to test different conditions at the same time. For more details refer to previously published methods on Western blots [18–20].

5.5.4 Troubleshooting

Western blot analysis may not always give good results. It is therefore very important in these cases to be able to identify problems and quickly find solutions. In Table 5.6, most common problems are listed along with some troubleshooting tips.

Table 5.6 Western Blot Common Problems and Troubleshooting Tips

Problems	Possible Cause	Solution
Weak signal	Not enough protein loaded on the gel	Check for total protein quantitation results accuracy
	Protein of interest weakly expressed in the sample	Load more proteins
		Try to concentrate protein samples
	Transfer problems	Verify transfer quality with Ponceau staining
		Check for protein retention in the gel following transfer (Coomassie staining)
		Use fresh transfer buffer
	Blocking agent too strong (antigens are masked)	Reduce blocking agent concentration
		Modify blocking step parameters (temperature and incubation time)
		Use another blocking agent
No signal	Transfer problems	Ensure completion of the transfer with Ponceau staining
		Verify that the transfer sandwich is well oriented in the electrotransfer chamber
	Primary antibody has low affinity for the antigen	Increase incubation time
		Optimize antibody concentration
		Use another antibody
	Wrong secondary antibody	Secondary antibody must be directed against the primary antibody species
	Incompatible or expired detection substrate	Use new detection substrate
		Check for compatibility with the secondary antibody conjugated enzyme
High background	Insufficient blocking	Use recommended concentration of blocking agent and incubation time
		Optimization of the blocking step
	Nonoptimal primary and secondary antibody concentration	Decrease antibody concentrations
		Optimize incubation conditions (time and temperature)
	Poor washing steps	Increase length and number of washes
		Add detergent to the washing buffer
	Membranes have dried out during the process	Keep membranes moist at all times
		Make sure sufficient volume is used to cover membranes
Unusual extra bands	Primary antibody cross-reactivity	Use another antibody
		Check in the literature
	Protein degradation (low molecular weight bands)	Use adequate homogenization buffer with protease inhibitors
		Always work on ice for sample preparation

5.5.5 Case Study

Studies on the potential toxic effects of Cd–tellurium-based nanoparticles (quantum dots) were undertaken to address the stability of quantum dots in biological systems. Indeed, quantum dots could liberate ionic Cd during breakdown in cells, which could lead to cytotoxicity. In addition, questions were asked if other properties (high surface area) of the nanoparticles could induce MTs by other mechanisms. To address these questions, rainbow trout hepatocytes were prepared and plated in increasing concentrations of uncoated

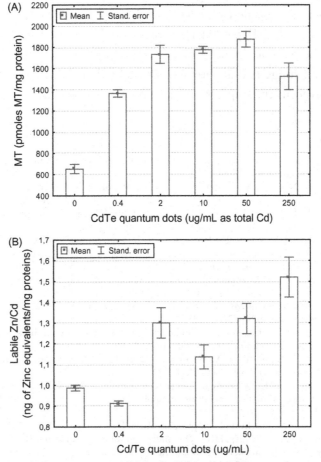

Figure 5.5 Changes in MT contents and labile Zn/Cd in rainbow trout hepatocytes exposed to quantum dots. Rainbow trout hepatocytes were exposed to an increasing concentration of quantum dots for 48 h at 15°C. The levels of MT (A) and labile Zn/Cd (B) were determined. The data represent the mean with SE.

quantum dots in culture media: 0, 0.4, 2, 10, 50, and 250 µg/mL). The cells
(1×10^6 cells/mL of culture media) were exposed for 48 h at 15°C, removed
from the culture media by centrifugation ($100 \times g$ 5 min), and resuspended in
PBS. The cells were analyzed for total MT levels and the levels of labile Zn/
Cd using the protocols described in this chapter.

The results show that the levels of MT in hepatocytes increased with the
added concentration of Cd-based quantum dots (Figure 5.5A). Indeed, MT
was significantly induced at 0.4 µg/mL of the total Cd in the exposure media.
The levels of labile Zn/Cd were also significantly elevated at a total Cd concentration of 2 µg/mL, suggesting the involvement of ionic Cd and Zn in cells.
Correlation analysis revealed that MT levels were significantly related to labile
Zn/Cd in cells ($r = 0.41$; $p = 0.01$). However, the induction of MT occurred
without any increase in labile Zn/Cd at 0.4 µg/mL, and the presence of a
weak correlation suggests that interactions other than the release of labile Cd^{2+}
might be involved in the expression of MT proteins in cells.

REFERENCES

[1] Shariati F, Shariati S. Review on methods for determination of metallothioneins in aquatic organisms. Biol Trace Elem Res 2011;141:340−66.
[2] Katakai K, Liu J, Nakajima K, Keefer LK, Waalkes MP. Nitric oxide induces metallothionein (MT)
gene expression apparently by displacing zinc bound to MT. Toxicol Lett 2011;119:103−8.
[3] Bauman JW, Liu J, Liu YP, Klaassen CD. Increase in metallothionein produced by chemicals that
induce oxidative stress. Toxic Appl Pharmacol 1991;110:347−54.
[4] Bonneris E, Perceval A, Masson S, Hare L, Campbell PGC. Sub-cellular partitioning of Cd, Cu and
Zn in tissues of indigenous unionid bivalves living along a metal exposure gradient and links to
metal-induced effects. Environ Pollut 2004;135:195−208.
[5] Gagné F, Blaise C. Available intracellular Zn as a potential indicator of heavy metal exposure in rainbow trout hepatocytes. Environ Toxicol Wat Qual 1996;11:319−25.
[6] Scheuhammer AM, Cherian MG. Quantification of metallothioneins by a silver-saturation method.
Toxicol Appl Pharmacol 1986;82:417−25.
[7] Gagné F, Marion M, Denizeau F. Metal homeostasis and metallothionein induction in rainbow trout
hepatocytes exposed to cadmium. Fund Appl Toxicol 1990;14:429−37.
[8] Ryvolova M, Krizkova S, Adam V, Beklova M, Trnkova L, et al. Analytical methods for metallothionein detection. Curr Anal Chem 2011;7:243−61.
[9] Viarengo A, Ponzanon E, Dondero F, Fabbri R. A simple spectrophotometric method for metallothionein evaluation in marine organisms: an application to Mediterranean and Antarctic molluscs.
Mar Environ Res 1997;44:69−84.
[10] Lindquist S, Craig EA. The heat-shock proteins. Annu Rev Genet 1988;22:631−77.
[11] Eckwert H, Alberti G, Kohler HR. The induction of stress proteins (hsp) in Oniscus asellus
(Isopoda) as a molecular marker of multiple heavy metal exposure: I. Principles and toxicological
assessment. Ecotoxicology 1997;6:249−62.
[12] Piano A, Valbonesi P, Fabbri E. Expression of cytoprotective proteins, heat shock protein 70 and
metallothioneins, in tissues of Ostrea edulis exposed to heat and heavy metals. Cell Stress
Chaperones 2004;9(2):134−42.

[13] Werner I, Nagel R. Stress proteins hsp60 and hsp70 in threes species of amphipods exposed to cadmium, diazinon, dieldrin, and fluoranthene. Environ Toxicol Chem 1997;16:2393−403.

[14] Sanders BM. Stress proteins in aquatic organisms: an environmental perspective. Crit Rev Toxicol 1993;23:49−75.

[15] Feder ME, Hofmann GE. Heat-shock proteins, molecular chaperones, and the stress response: evolutionary and ecological physiology. Annu Rev Physiol 1999;61:243−82.

[16] Towbin H, Staehelin T, Gordon J. Electrophoretic transfer of proteins from polyacrylamide gels to nitrocellulose sheets: procedure and some applications. Proc Natl Acad Sci USA 1979;76:4350−4.

[17] Romero-Calvo I, Ocón B, Martínez-Moya P, Suárez MD, Zarzuelo A, et al. Reversible Ponceau staining as a loading control alternative to actin in Western blots. Anal Biochem 2010;401 (2):318−20.

[18] Gallagher S, Winston SE, Fuller SA, Hurrell JG. Immunoblotting and immunodetection. Curr Protoc Cell Biol 2011; 6.2.1−6.2.28.

[19] Gassmann M, Grenacher B, Rohde B, Vogel J. Quantifying Western blots: pitfalls of densitometry. Electrophoresis 2009;30:1845−55.

[20] Kurien BT, Scofield RH. Western blotting. Methods 2006;38:283−93.

Oxidative Stress

François Gagné

Chapter Outline

Oxidative stress represents a common if not universal denominator of toxicity that could lead to tissue injury and threaten organism health. During respiration, the release of uncoupled high-energy electrons by mitochondria and chloroplasts represents one of the major sources of reactive oxygen species in cells. Many xenobiotics have the potential to uncouple electron transport in mitochondria, which could lead to important leakage of electrons in the intracellular environment and produce oxidative damage. The production of reactive oxygen species could also result from inflammation reactions in which nitric oxide and hydrogen peroxide are produced and during oxidative metabolism of xenobiotics. We invite the reader to consult recent reviews on oxidative stress in the environment for a more detailed account on the role of oxidative stress during pathogenesis [1].

Oxidative stress occurs when excess oxygen radicals are produced in cells, which could overwhelm the normal antioxidant capacity. When the concentration of reactive species is not controlled by internal defense mechanisms such as antioxidants (tocopherols, ascorbic acid, and glutathione) or enzymes involved in oxygen radical scavenging (catalase, peroxidase, and superoxide dismutase, SOD), oxidative damage occurs to proteins, lipids, and DNA, which could lead to cytotoxicity, genotoxicity, and even carcinogenesis when damaged (mutated) cells can proliferate. Oxidative stress could results from the following: (1) the presence of xenobiotics, (2) the activation of the immune system in response to invading microorganisms (inflammation), and (3) radiation, which makes oxidative stress a common denominator of toxicity or stress. In this chapter, the reader will find a variety of assays to measure oxidative stress and damage.

6.1 ANTIOXIDANT ENZYMES

6.1.1 SOD

SODs are found in both the mitochondria and cytoplasm in cells where the latter is dependent of Cu and Zn. SOD catalyzes the dismutation of oxygen radicals to produce hydrogen peroxide, which in turn is transformed into water and oxygen by catalase:

$$\text{Superoxide anion} \rightarrow H_2O_2(\text{SOD})$$
$$H_2O_2 \rightarrow O_2 + H_2O \text{ (catalase)}$$

In the present assay, a nonenzymatic means of superoxide generation is presented in which the generated radicals are detected by the reduction of the colorimetric dye p-iodonitrotetrazolium (p-INT) [2]. This method could be run on 96-well microplates for high-throughput analysis.

6.1.1.1 Reagents

Homogenization buffer: 50-140 mM NaCl containing 10 mM HEPES-NaOH, pH 7.4, containing a protease inhibitor (phenylmethylsulfonylfluoride 0.01% or apoprotinin 1 µg/mL) and a reducing agent (5 mM β-mercaptoethanol or 5 mM dithiothreitol).

Assay buffer: 50 mM KH_2PO_4 at pH 7.5 with NaOH 1 M: add 0.68 g KH_2PO_4 in 90 mL of SQ water, adjust pH to 7.5, and complete to 100 mL SQ water. Keep at 4°C.

Stock EDTA (10×): 9 mg of EDTA disodium salt in 10 mL water.

NADH stock: Resuspend 9.5 mg of reduced NADH in 1.4 mL of water. Prepare daily.

p-INT stock (5 mM): Dissolve 25 mg of p-INT in 10 mL of SQ water. Keep in the dark at 4°C, stable for one week.

Phenazine methosulfate (PMS) stock solution: Dissolve 6 mg in 20 mL of SQ water. Keep at −20°C. If the clear liquid becomes brown then discard.

SOD enzyme concentrate: Dissolve commercial SOD (4000 U/mg solid) in 10 mL of assay to obtain 1000 U/mL. Conserve 100 µL aliquot at −20°C for up to 2 years. Dilute again 1/10 in the assay buffer on the day of the assays.

Reagent 1: Prepare at the day of the assay the following: 165 µL of PMS and 250 µL of EDTA stock and complete to 5 mL with assay buffer.

Reagent 2: 1 mL of EDTA, 3.2 mL of p-INT stock, and 1.96 mL of NADH and complete to 10 mL with the assay buffer.

6.1.1.2 Procedure

The tissues are weighed and minced with a scalpel on ice. The minced tissues are homogenized by using a Teflon pestle tissue grinder and the homogenate centrifuged at 15,000 × g for 30 min at 2−4°C. The supernatant (S15) is kept aside for the assay.

In a clear polystyrene 96-well microplate, add 50−100 µL of S15, blank, or SOD sample to an equal volume of Reagent 2. Start the reaction by adding 25 µL of Reagent 1. Determine the absorbance at 560 nm at time 0, 2, 4, 8, and 16 min. The SOD sample contains 10 U/mL and is prepared as follows: 22 µL of SOD stock solution and 88 µL of assay buffer (100 µL). Continue as with the S15 sample by adding 1 volume of Reagent 2 and 25 µL of Reagent 1. The blank solution contains only the assay buffer.

6.1.1.3 Data Calculation and Normalization

The presence of SOD will block the reduction of the dye by the electron generation system. The decrease in absorbance is determined over time and the slope is calculated to have Δ absorbance/min. The percentage of inhibition by SOD is first determined as follows:

$$\frac{[(\Delta A560\ nm/min_{Blank} - (\Delta A560\ nm/min_{SOD})] \times 100 = \%inhibition,}{(\Delta A560\ nm/min_{Blank})}$$

and the percentage of inhibition of the S15 sample determined:

$$\frac{[(\Delta A560\ nm/min_{Blank} - (\Delta A560\ nm/min_{S15})] \times 100 = \%inhibition}{(\Delta A560\ nm/min_{Blank})}$$

The percentage of inhibition in the S15 fraction is converted into SOD equivalents by the following:

$$SOD\ (U/mL) = (x\%\ inhibition\ S15/y\%\ inhibition\ of\ SOD) \times 10\ U/mL$$

The SOD activity is then normalized by the amount of proteins in the S15 fraction or by the tissue weight to give SOD U/mg proteins or g tissue weight.

6.1.2 Peroxidase

Peroxidase is a hemoprotein catalyzing the oxidation of a number of substrates by hydrogen peroxide:

$$H_2O_2 + substrate - H2 \rightarrow substrate + 2\ H_2O$$

Endogenous substrates are, for example, ascorbate, ferrocyanide, Fe^{3+}-cytochrome c, and reduced glutathione (2 GSH). The enzyme requires a substrate susceptible to oxidation and represents a fundamental pathway of oxidation of many xenobiotics. In the present assay, a highly sensitive fluorescence measurement of fluorescein is proposed as a convenient procedure. For more sensitivity, luminol (3-aminophthalhydrazide) could be also used as luminescent assay if a luminescence reading instrument is available.

6.1.2.1 Reagents

Assay buffer: 50 mM KH_2PO_4 at pH 7.5 with NaOH 1 M: add 0.68 g KH_2PO_4 in 90 mL of SQ water, adjust pH to 7.2, and complete to 100 mL SQ water. Keep at 4°C.

Dichlorofluorescein diacetate 10 mM (DCF): Dissolve 48.5 mg DCF in 10 mL 100% dimethylsulfoxide.

Horseradish peroxidase (HRP): 1 mg/mL in 0.1 M KH_2PO_4 buffer, pH 6. The buffer is prepared by dissolving 1.36 g of monobasic potassium phosphate in 90 mL water, pH adjusted to 6 with HCl, and complete to 100 mL with SQ water. A volume of 10 µL of the stock is added to 100 mL of assay buffer to produce 0.1 µg/mL.

Reagent solution: Mix 100 µL of DCF and 100 µL of H_2O_2 at 1 mM concentration and complete to 10 mL with assay buffer. Prepare just before the assays and keep on ice in the dark.

Fluorescein standard solution (100 µg/mL): Dissolve 1 mg fluorescein in 10 mL of assay buffer.

6.1.2.2 Procedure

In a dark 96-well microplate, add 20 µL of S15, standard, or blank to 160 µL reagent solution and 20 µL of water. Incubate at 25–30°C and measure fluorescence at 485 nm excitation and 520 nm emission wavelengths at time 0, 5, 10, 15, and 20 min. The blank consists of the assay buffer only. A positive internal control using 10 ng/mL peroxidase is added to the S15 to express the data as peroxidase equivalents (standard addition method). A volume 20 µL of the diluted stock solution is added to 180 µL reagent solution with S15 (final peroxidase concentration 10 ng/mL). The activity could also be expressed as nmol fluorescein formed by preparing an external 1 µg/mL fluorescein standard: 2 µL of fluorescein standard solution to 198 µL of assay buffer.

6.1.2.3 Data Calculation and Normalization

Peroxidase activity (ng/mL) = [RFU (sample − blank)/min]/[RFU (spiked sample − blank)/min] × added peroxidase (ng/mL) × dilution factor

The equivalent amount of peroxidase is then normalized against total protein content in the S15 fraction or the tissue weight (g/mL homogenization buffer) to give ng peroxidase/mg proteins.

6.1.3 Catalase

Catalase is the enzyme responsible for the breakdown of H_2O_2 into water and oxygen. It is the major antioxidant defense enzyme system. Catalase is inhibited by aminotriazole at concentrations between 1 − 10 mM.

$$2\,H_2O_2 \rightarrow 2\,H_2O + O_2$$

The assay is based on the time-dependent elimination of H_2O_2, which is determined by the DCF substrate described earlier.

6.1.3.1 Reagents

Assay buffer: 100 mM KH_2PO_4, pH 7.0. Dissolve 1.36 g KH_2PO_4 in 90 mL bidistilled water, adjust pH to 7 with NaOH 1 M, and complete to 100 mL with bidistilled water.

Hydrogen peroxide reagent: Prepare a 100 mM (or 0.2%) solution in bidistilled water. Hydrogen peroxide is usually sold at $\sim$30% concentrate. Prepare daily and keep at 4°C.

Peroxidase reagent: Dissolve 10 mg HRP and 1 mg of DCF in 100 mL of assay buffer. Dilute 1/10 in the assay buffer.

Fluorescein standard (10 μM): Dissolve 3 mg in 10 mL of bidistilled water. Dilute 1/100 in assay buffer to obtain the final concentration 3 μg/mL.

6.1.3.2 Procedure

Catalase is a cytosolic enzyme and the assay is commonly practiced on the post-mitochondrial fraction ($>$ 10,000 $\times g$ supernatant of tissue homogenates). Mix 10−50 μL of 10 to 15 000 $\times g$ supernatant, blank (homogenization buffer), or hydrogen peroxide standard (1−10 mM H_2O_2) and complete to 180 μL with the assay buffer. Start the reaction by adding 20 μL of hydrogen peroxide reagent and incubate for 0, 10, 20, 30, and 40 min at 20−22°C.

At each incubation time, pipette 20 μL of reaction mixture and mix with 80 μL of peroxidase reagent and incubate for 15 min and measure fluorescence at 485 nm excitation and 520 nm emission. Blanks consist of 20 μL of bidistilled water and a fluorescein standard of 1 μM could be used for calibration (10 μL of fluorescein standard in 80 μL of the assay buffer). The amount of H_2O_2 will decrease over time by catalase.

6.1.3.3 Data Calculation

Catalase activity will eliminate hydrogen peroxide, which in turn will reduce the amount of fluorescein (less oxidation of the DCF substrate). It is calculated as follows:

$$[(\text{Fluorescein formed}/15\,\text{min})\,\text{at}\,0\,\text{min}] - [(\text{Fluorescein formed}/15\,\text{min})\,\text{at}\,20\,\text{min}]$$
$$= \text{loss of hydrogen peroxide}/\text{min}/\text{mL}$$

This activity is then normalized against total protein content (mg/mL) or tissue weight (g/mL) to give loss of hydrogen peroxide/min/mg proteins or g tissue weight.

6.1.4 Measurement of Antioxidant Capacity

During oxidative stress, reactive oxygen species reach levels that could overwhelm antioxidant defense in cells; a decrease in the levels of antioxidants such as ascorbic acid or reduced glutathione levels leads to general reduction of the antioxidant capacity of the cytoplasm. An extremely cheap and rapid method to determine the antioxidant capacity in tissue extracts is presented therein. The assay consists of the reduction of the phosphomolybdate complex by the presence of antioxidants in the biological sample [3,4]. This complex forms a dark blue chromophore when reduced that is measured between 660 and 850 nm with a maximum at 810 nm. This assay requires only a standard spectrophotometer for tubes or microplates. These data are usually expressed as ascorbic acid equivalents, which are used as positive controls.

6.1.4.1 Reagents and Tissue Preparation

Tissues are usually homogenized in a buffer that is devoid of any antioxidant and the buffer should be used in the blank. Trichloroacetic acid (TCA) is added to the homogenate sample to give a final concentration of 5% (w/v) and placed in ice. The mixture is then centrifuged at $10,000 \times g$ for 5 min at 4°C and the supernatant collected.

Phosphomolybdate reagent: Dissolve 10 mg phosphomolybdate [$H_3(P(Mo_3)O_{10})_4$] in 1 mL of 20% ethanol (10 mg/mL) in a dark container and store at 4°C. Dilute at 50 μg/mL in 0.5% TCA before the assays.

Ascorbic acid reagent: Dissolve 10 mg of ascorbic acid in 1 mL of SQ water. Prepare fresh (daily) and place on ice in the dark to prevent oxidation (maximum concentration).

6.1.4.2 Procedure

To 50 μL of blank, ascorbic acid standards, and tissue homogenate acid extract mix 150 μL of phosphomolybdate reagent and wait for 15 min at room temperature. Read absorbance at 810 nm. The blank is 5% TCA in the homogenization buffer.

6.1.4.3 Data Calculation, Normalization, and Comments

[A810 sample − A810 blank] × Standard concentration of ascorbic acid/

[A810 standard − A810 blank] × (sample volume μL/total assay volume μL)

= Y mg/mL

This concentration is then normalized against total proteins (mg/mL) or homogenate tissue weight in g/mL:

Y μg/mL/g tissue weight/mL or mg/mL protein = Z μg/g or Z μg/mg proteins).

The concentration of the standard solution should be close to the signal value of the sample, i.e., within ± 0.2 absorbance unit. A standard curve could also be constructed to convert the increase in absorbance at 810 nm as function of ascorbic acid concentration. The concentration of ascorbic acid is usually between 1 and 10 mg/mL. If the test sample significantly absorbs at 810 nm then an operational correction blank could be used, which consists of 50 μL of the sample with 150 μL of 0.5% TCA only. These data are expressed as mg ascorbic acid equivalents/mg proteins.

6.1.5 Lipid Peroxidation (Oxidative Damage)

Excess reactive oxygen radicals that are not removed by antioxidants or catalase could lead to deleterious oxidative reactions in cells. The carbon double bonds in polyunsaturated lipids could react with hydrogen peroxides to give the breakdown product malonaldehyde (MDA). The thiobarbituric test represents a very convenient and simple assay for the measurement of lipid peroxidation [5]. MDA is detected by the thiobarbituric acid, which forms a chromophore detectable by either spectrophotometry or fluorometry. Therefore this assay represents the extent of breakdown products resulting from the oxidation of polyunsaturated lipids and hence a biomarker of oxidative damage. The assay is usually practiced on lipid-rich samples such as tissue homogenates and mitochondrial and microsomal fractions.

6.1.5.1 Reagents

Reagent A: Add 10 mL of trichloroacetic acid (6.1 M) to 70 mL of water, dissolve 0.028 g of $FeSO_4$, and complete to 100 mL with water;

Reagent B: Add 0.67 g of thiobarbituric acid to 100 mL of SQ water and heat at 50−60°C to assist dissolution;

Tetramethoxypropane (TMP) standard: TMP is the stabilized formulation of MDA. A standard stock solution is prepared by adding 10 µL of TMP in 10 mL of 0.1 M HCl. This stock solution is stable for one week at 4°C in the dark. The working solution 10 µg/mL is freshly prepared before the assay by mixing 10 µL of stock solution in 1 mL SQ water.

6.1.5.2 Procedure

In a dark 96-well microplate, add 150 µL of Reagent A and 75 µL of Reagent B to 50 µL of blank, standard, or sample. Mix, heat in water bath at 70–75°C for 10 min, and cool at room temperature. The microplate could be centrifuged briefly if there is precipitation (protein denaturation) at $3000 \times g$ for 5 min. Read fluorescence at 535 nm excitation and 635 nm emission or read the absorbance at 540 nm.

6.1.5.3 Data Calculation, Normalization, and Comments

[A540 sample − A540 blank] × Standard concentration/

[A540 standard − A540 blank] × (sample volume µL/total assay volume µL)

$= Y$ µg/mL

This concentration is then normalized against total proteins (mg/mL) or homogenate tissue weight in g/mL:

Y µg/mL/g tissue weight/mL or mg/mL protein $= Z$ µg/g or Z µg/mg proteins).

The concentration of the standard solution should be close to the signal value of the sample, i.e., within ± 0.2 absorbance units or $\pm 20\%$ of the fluorescence signal. A standard curve could also be constructed to convert the absorbance values into MDA equivalents. Since other aldehydes could react with the reagent, these data are usually expressed as µg thiobarbituric acid reactants (TBARS).

6.1.6 Measurement of Age-Related Pigments and Lipofuscins

Age-related pigments such as lipofuscins represent a simple way to determine the rate of living or the physiological age of organisms living in polluted environments [6]. Age-related pigments are formed by the chronic oxidation of proteins, lipids, and carbohydrates in cells. These pigments form insoluble inclusion bodies and are associated with neutral lipids and lipofuscins during histocytochemical observations [7]. These inclusions are fluorescent and accumulate over time and persist in cells (no turnover) that can be used to assess the "physiological" age of organisms provided these inclusions are related with chronological age during normal conditions. This needs to be checked when working with novel tissues as in reference [6] where a correlation between age-related pigments and chronological age was found at reference sites.

6.1.6.1 Reagents and Solutions

Homogenization buffer: 25 mM HEPES-NaOH buffer, pH 7.4, containing 125 mM NaCl, 10 µg/mL apoprotinin, and 1 mM dithiothreitol.

Ethanol: Analytical grade absolute ethanol (99—100%), 100 mL.

Extraction buffer: Prepare 10 mM sodium phosphate buffer, pH 7.4, containing 125 mM NaCl and 0.5% Tween 20.

Quinine sulfate: Prepare a stock solution of quinine sulfate at 1 mg/mL in 100% ethanol. Dilute at 0.1 µg/mL in 0.05 N H_2SO_4.

6.1.6.2 Procedure

The levels of age-related pigments in the foot of the mussel were measured according to the fluorescence methodology [8]. Age-related pigments are usually performed on tissues with low turnover such as the skin, mantle, or foot in mussels or clams. Tissues are homogenized at a 1:5 ratio in the homogenization buffer using a Polytron or Teflon pestle tissue grinder at 2—4°C. The homogenate was decanted to remove tissue clumps and 2 volumes of absolute ethanol were mixed to 1 volume of homogenate for 10 min at room temperature. The mixture was then centrifuged at $10,000 \times g$ for 5 min to precipitate denatured proteins. The supernatant is referred to as lipid–like and the pellet as protein–associated age-related pigments, respectively. The protein pellet was mixed with 2—3 volumes of the extraction buffer for 15 min and recentrifuged at $10,000 \times g$ for 5 min at room temperature. The supernatants were diluted 1:5 in either absolute ethanol or the extraction buffer. Fluorescence was determined at 350 and 460 nm excitation and emission, respectively. Quinine sulfate at the concentration 100 ng/mL was used as an internal standard for normalization of the lamp excitation energy, which can change in time and between instruments. Total proteins are also determined using the method of Bradford as described in Chapter 8.

6.1.6.3 Data Calculation

The same concentration of quinine should give the same fluorescence units within ± 20% of the measured value. If the value differs from this then a correction factor should be applied to sample batch showing higher/lower values in fluorescence units. For example, if the quinine fluorescence values for batches 1 and 2 are 100 and 85 fluorescence units then no correction is applied, because they are within 20% tolerance interval. If a third batch of sample gives a fluorescence value of 135 then the fluorescence values should be corrected against the mean fluorescence value of batches 1 and 2: The mean value of batches 1 and 2 is 92.5 fluorescence units and a correction factor is derived, 135/92.5 = 1.459. The fluorescence units of the samples in batch 3 should be divided by 1.459 before comparing with the values obtained for batches 1 and 2. Some investigators just apply the correction factor regardless of the ± 20% rule between sample batches.

The data are expressed as corrected relative fluorescence units/ mL × dilution factor [5] and normalized against total proteins (mg/mL) or g tissue wet or dry weight.

6.2 APPLICATIONS

The occurrence of pharmaceutical compounds in treated municipal waste-waters has raised concerns about their potential for harming nontarget aquatic organisms. Indeed, pharmaceuticals such as carbamazepine are designed to have specific therapeutic effects and to be eliminated either directly or indirectly by mammalian-based biotransformation activity. However, aquatic organisms such as the hydra may not have the biotransformation capacity to biotransform them for elimination (see Figure 6.1). In the example provided below, exposure of the Hydra to increasing concentrations of carbamazepine for 48 hr at room temperature. In Figure 6.1A, heme oxidase activity (a generic assay of cytochrome P450 activity) was significantly increased at 6 μM concentration. In Figure 6.1B, lipid peroxidation was significantly induced at 600 μM which indicates oxidative stress–induced cell damage.

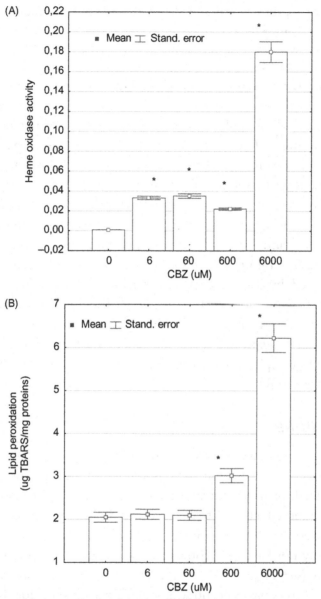

Figure 6.1 Increase in heme oxidase and lipid peroxidation in hydra exposed to carbamazepine. The activity of heme oxidase is involved in xenobiotic metabolism (A) and lipid peroxidation is a biomarker of oxidative damage to polyunsaturated lipids (B).

REFERENCES

[1] Lushchak V. Review: environmentally induced oxidative stress in aquatic animals. Aquatic Toxicol 2011;101:13—30.

[2] Ewing JF, Janero DR. Microplate superoxide dismutase assay employing a nonenzymatic superoxide generator. Anal Biochem 1995;232:243—8.

[3] Verlecar XN, Jena KB, Chainy GBN. Seasonal variation of oxidative biomarkers in gills and digestive gland of green-lipped mussel Perna viridis from Arabian Sea Estuarine. Coastal Shelf Sci 2008;76:745—52.

[4] Mitusi A, Ohata T. Photooxidative consumption and photoreductive formation of ascorbic acid in green leaves. Plant Cell Physiol 1961;2:31—44.

[5] Wilbur KM, Bernheim F, Shapiro OW. The thiobarbituric acid reagent as a test for the oxidation of unsaturated fatty acids by various agents. Arch Biochem Biophys 1949;24:305—13.

[6] Gagné F, Blaise C, Pellerin J, Fournier M, Gagnon C, Sherry J, et al. Impacts of pollution in feral *Mya arenaria* populations: the effects of clam bed distance from the shore. STOTEN 2009;407:5844—54.

[7] Sheehy MR, Greenwood JG, Fielder DR. Lipofuscin as a record of "rate of living" in an aquatic poikilotherm. J Gerontol A Biol Sci Med Sci 1995;50:327—36.

[8] Hammer C, Braun E. Quantification of age pigments (lipofuscin). Comp Biochem Physiol 1988;90B:7—17.

REFERENCES

CHAPTER 7

Xenobiotic Biotransformation

François Gagné

Chapter Outline

7.1 XENOBIOTIC BIOTRANSFORMATION

Over time, living organisms have developed the capacity to eliminate unwanted chemicals, so-called xenobiotics, from cells in order to maintain homeostasis.

Biochemical Ecotoxicology
DOI: http://dx.doi.org/10.1016/B978-0-12-411604-7.00007-6
117

Some xenobiotics are poorly soluble in aqueous media and tend to partition in membranes and lipid bodies. Organisms have developed a system of membrane bound hemoproteins capable of oxidizing nonpolar compounds such as polyaromatic and aliphatic hydrocarbons involving the addition of a single oxygen (mono-oxygenation reactions). Cytochrome P450 enzymes are a large family of heme-containing proteins involved in the hydroxylation (oxidation) of both endogenous and exogenous compounds such as steroids, phenobarbital, and polycyclic aromatic hydrocarbons. They are one of the most studied membrane protein families in biochemistry, pharmacology, and toxicology. These protein were initially found by the appearance of a chromophore at 450 nm upon carbon monoxide binding under reducing conditions [1]. This spectral characteristic led to its name, cytochrome P450. Cytochrome P450s are responsible for the phase 1 biotransformation of nonpolar lipophilic xenobiotics, which consist of the mono-oxygenation of the hydrocarbon backbone. Phase 1 reactions principally involve oxidation reactions (hydroxylations), but other reactions also could occur such as reduction, hydrolysis, and acetylation [2]. These cytochromes are regulated by the nuclear hormone receptor superfamily, which includes peroxisome proliferation-activated receptor or the pregnane X receptor (PXR) for various polycyclic aliphatic and heterocyclic compounds. The Ah receptor is activated by coplanar aromatic polycyclic compounds (dioxins, benzo(a)pyrene) and is from the PAS (for period circadian protein/aryl hydrocarbon nuclear translocator/single-minded proteins) receptor family.

Phase 2 reactions comprise a more diverse group of enzymes involved in the addition of polar compounds to hydroxylated xenobiotics or directly to the xenobiotics if polar groups are present. The most common reactions are the addition of glutathione (glutathione S-transferase, GST), of sulfate (sulfotransferase), and of glucuronyl sugars (uridine-5-diphosphate glucuronyl transferase). The active removal of potentially toxic compounds from cells (efflux) is performed by P-glycoprotein pumps located at the cytoplasmic membrane. While some people call this step a phase 3 biotransformation reaction, this multi-xenobiotic resistance mechanism is an elimination mechanism instead of a biotransformation mechanism. The reader will find the most common methods for the evaluation of xenobiotic biotransformation in organisms in this chapter.

7.2 GENERIC ASSAY FOR CYTOCHROME P450

Cytochrome P450-related activities are usually determined by using specific substrates that target particular isoforms of cytochrome P450 (1A1, 1A2, 3A4, etc.), enzyme-linked immunoassays and targeted gene expression for mRNA. However, the investigator is sometimes confronted with new organisms from which no information about gene/protein sequences and related enzyme activity are available. In this case, a generic assay for cytochrome P450 activity based on fundamental biochemical properties is proposed. The basic properties of any hemoproteins consist in their ability to act as nonspecific peroxidases when denatured. Thus, the principle of this assay resides in the capacity of denatured hemoproteins in microsomes (composed of rough and smooth endoplasmic reticulum) to show peroxidase activity. Hence, the evaluation of peroxidase using nonspecific substrates in experimentally denatured samples represents a fundamental, convenient, and cheap means to measure the global hemoprotein activity in microsomes that abound in cytochrome P450s. This method was originally developed to stain cytochrome P450 bands in denaturing gel electrophoresis [3] and was modified for spectrophotometric assessment. The assay is simple, cheap, and easy to implement. It is usually practiced in microsomal preparations where hemoproteins are mostly cytochrome P450, but it can be practiced in the post-mitochondrial fraction to determine the hemoprotein peroxidase activity before and after denaturation by mild acid and detergent treatments. Background peroxidase, which is found in cytosol, is measured before denaturation and cytochrome P450 activity after denaturation or after filtrating the supernatant on a 0.22 um membrane.

7.2.1 Reagents

Homogenization buffer: 10 mM HEPES–NaOH, pH 7.4, containing 1.15% KCl, 1 mM EDTA, and 1 µg/mL protease inhibitor (apoprotinin) or 0.1−1 mM phenylmethanesulfonyl fluoride (PMSF).

Denaturing buffer: Prepare a 0.2 M sodium acetate buffer, pH 5, containing 0.1% (w/v) Triton X-100.

Benzidine substrate: Prepare a stock solution at a concentration of 250 mM in 95% ethanol. Benzidine is considered carcinogenic and proper safety precautions should be taken: gloves, security glasses, and work under a fume hood. Other substrates could be used such as diaminobenzidine or 2,2′-azino-di (3-ethylbenzthiazoline-6-sulfonate, ABTS; at 405 nm).

Hydrogen peroxide solution: Prepare a 100 mM (0.2%) working solution in bidistilled water. Prepare daily. Keep on ice (4°C).

7.2.2 Procedure

For cytochrome P450, the microsomal fraction should be isolated. Homogenize tissue/cell samples in ice-cold homogenization buffer with a Teflon pestle tissue grinder (five passes). Centrifuge at $9-12,000 \times g$ for 20 min at $2-4°C$ and carefully remove the supernatant from the pellet and upper lipid layer. Microsomes could be isolated at $100,000 \times g$ for 60 min if an ultracentrifuge is available. If no ultracentrifuge is available then the microsomes could be prepared by the calcium aggregation method [4,5]. Calcium is added to the supernatant to reach a final concentration of 10 mM $CaCl_2$, incubated for 10 min on ice, and centrifuged at $15,000 \times g$ for 30 min at $4°C$. The pellet contains the microsomes and is resuspended in 2 volumes of the homogenization buffer.

For the peroxidase activity assay, mix in a clear microplate or tube consisting of 50 μL of microsomal preparation or water (blank), 130 μL of denaturing buffer (pre-incubate for 15 min on ice), start the reaction by adding 10 μL of benzidine substrate and 10 μL of hydrogen peroxide. Incubate for $30-60$ min at $30°C$ and determine absorbance at 650 nm.

7.2.3 Data Calculation

Report the enzyme activity in Absorbance change $/(\min \times mL)$ with the following:

$$[(A650 \text{ sample} - A650 \text{ blank})/30 \text{ or } 60 \text{ min}] \times (\text{dilution factor } 200/50) \times$$
$$1/\text{volume used mL} = \text{Absorbance change}/\min/mL.$$

The enzyme activity is then normalized by the protein concentration of the microsomal preparation or tissue wet or dry weight:

$$\text{Enzyme activity}/\text{mg proteins}/mL \text{ or tissue weight of homogenate g}/mL$$
$$= \text{Change in absorbance}/\min/\text{mg proteins or g tissue.}$$

7.3 CYTOCHROME P4501A1 AND BENZO(A)PYRENE HYDROXYLASE ACTIVITIES

The activity of cytochrome P450A1A1 is responsible for benzo(a)pyrene hydroxylase activity, which represents the standard assay for Ah receptor-mediated induction of cytochrome P4501A1 activity [6,7]. However, this assay manipulates benzo(a)pyrene, a potent carcinogenic compound, which is a health risk for laboratory personnel and was soon replaced by the safer 7-ethoxyresorufin substrate, which is O-dealkylated to produce fluorescent 7-hydroxyresorufin by cytochrome P4501A1 [8]. This substrate is safe and gives a relatively sensitive fluorescent signal, which is amenable to cost-effective and high-throughput screening if a fluorescent microplate reader is available.

7.3.1 Reagents

Homogenization buffer: Prepare 25 mM HEPES-NaOH, pH 7.4, containing 1.15% KCl, 1 mM EDTA, and 0.1 mM PMSF.

Assay buffer: Prepare 100 mM potassium phosphate buffer, pH 7.4.

Substrate mix: Prepare in separate microcentrifuge tubes 1 mM NAPDH (25 mg of reduced NADPH in 22 mL of assay buffer, prepare fresh) and 100 μM of 7-ethoxyresorufin in the assay buffer. A concentrate stock solution at 10 mM could be prepared in dimethylsulfoxide (DMSO): dissolve 5 mg 7-ethoxyresorufin in 2.1 mL of DMSO 100%, store in aliquots in the dark at −20°C, and dilute 1/100 with the assay buffer at the day of the assay.

Standard: Prepare 100 μM of 7-hydroxyresorufin working solution daily. A stock solution of 10 mM could be prepared in the assay buffer and stored at 4°C for 6 months in the dark.

7.3.2 Procedure

The assay is usually practiced on the post-mitochondrial fraction, i.e., the supernatant obtained between 9 and 15,000 × g centrifugation. The tissues are usually homogenized at 20% tissue weight/homogenization buffer volume ratio in iso-osmotic buffer such as the one described previously to preserve membrane integrity. In a dark microplate (fluorescence), mix the following in duplicate and separate wells: 50 μL of post-mitochondrial fraction, water (blank), or standard; 20 μL of substrate; 20 μL of NAPDH; and 90 μL of assay buffer. Incubate at 30°C and read fluorescence at each 10 min interval for a maximum of 1 h at 540 nm excitation and 600 nm emission using a microplate fluorescence reader. If no microplate readers are available, the samples are read at the end of incubation time (30 min or 60 min) by adding 1 mL of 1 mM NaOH (stops the reaction by dilution and high pH) to the reaction mixture and measured using a standard cuvette fluorescence spectrometer. Take the readings immediately at the same wavelengths described previously using 10−20 nm bandpasses.

7.3.3 Data Calculation

The rate of dealkylation is determined as follows:

$$(RFUsample - RFUblank) \times \text{Standard concentration (25 nmol/mL)}$$
$$/(RFUstandard - RFUblank) \times 200/50 \times 1/\text{time in min},$$

where RFUs are relative fluorescence units, 200/50 the dilution factor of the assay, and the standard concentration is 25 nmol/mL 7-hydroxyresorufin.

The data are then normalized against total protein levels in the supernatant or tissue weight/mL to give the enzyme activity: nmol resorufin formed/min/mg proteins.

7.4 CYTOCHROME P4503A4 ACTIVITY

The activity of cytochrome P450A3A4 was recently proposed as a marker of exposure to pharmaceutical drugs or hydroxylated polycyclic aliphatic and aromatic hydrocarbons in aquatic organisms [9] and in mammals [10]. The assay is based on the O-dealkylation of dibenzyloxyfluorescein (DBF) to give fluorescein, which is a highly sensitive fluorescent molecule (high quantum yield). This enzyme is also responsible for testosterone 6β hydroxylase activity, which represents a catabolic/inactivation pathway for this sex steroid.

7.4.1 Reagents

Homogenization buffer: Prepare 25 mM HEPES-NaOH, pH 7.4, containing 1.15% KCl, 1 mM EDTA, and 0.1 mM PMSF.

Assay buffer: Prepare 50 mM Tris-HCl, pH 7.4, containing 0.05% Tween 20.

Substrate mix: Prepare in separate microcentrifuge tubes 1 mM NAPDH (25 mg of reduced NADPH/22 mL of assay buffer, prepare fresh) and 10 μM DBF in the assay buffer (10 μL of stock solution in 10 mL of assay buffer). A concentrate stock solution at 10 mM could be prepared in acetonitrile: dissolve 9.7 mg of DBF in 970 μL of acetonitrile 100%, and store in aliquots in the dark at −20°C.

Standard: Prepare 10 μM of fluorescein working solution daily. A stock solution of 10 mM could be prepared in the assay buffer and stored at 4°C for 6 months in the dark.

7.4.2 Procedure

The assay is usually practiced on the post-mitochondrial fraction supernatant at >9–15 000 $\times g$ centrifugation for 20-30 min at 4°C. Tissues are usually homogenized at 20% tissue weight/homogenization buffer volume ratio in iso-osmotic buffers such as the one described above. In a dark microplate (fluorescence) mix the following in duplicate: 50 μL of post-mitochondrial fraction, water (blank), or standard; 20 μL of substrate; 20 μL of NAPDH; and 90 μL of assay buffer. Incubate at 30°C and read fluorescence at each 10 min interval for up to 1 h at 485 nm excitation and 520 nm emission using a microplate fluorescence reader. If no microplate readers are available, the samples could be measured at the end of incubation time (30 min or 60 min) by adding 1 mL of 1 mM NaOH (stops the reaction by dilution and high pH) to the reaction mixture and measured using a standard cuvette fluorescence spectrometer. Take the readings immediately.

7.4.3 Data Calculation

The rate of dealkylation is determined as follows:

$$(RFUsample - RFUblank) \times Standard\ concentration(2.5\ nmol/mL)$$
$$/RFUstandard - RFUblank \times 200/50 \times 1/min,$$

where 200/50 the dilution factor of the assay and the standard concentration is 2.5 nmol/mL fluorescein. These data are then normalized against total protein levels in the supernatant or tissue weight/mL to give the enzyme activity: nmol resorufin formed/min/mg proteins.

7.5 GLUTATHIONE S-TRANSFERASE ACTIVITY

The conjugation of glutathione to polar xenobiotics by GST represents a major pathway for phase 2 reactions in fish and especially in invertebrates. For example, tissue burden in heavy polyaromatic hydrocarbons such as benzo(a)pyrene was significantly correlated with GST activity in wild mussel populations [11] providing evidence along with other investigations of its involvement in biotransformation. This method is based on the conjugation reaction of reduced glutathione with 1-chloro-2-4-dinitrobenzene substrate, which is broadly specific to the various isoforms of the enzyme [12,13]. This assay is cheap, easy to perform, and could be very rapid if a microplate reader (absorbance) is available.

7.5.1 Reagents

Assay buffer: Prepare 100 mL of 125 mM NaCl and 10 mM HEPES-NaOH, pH 6.5.

Reduced GSH reagent: Prepare 1 mM of reduced GSH in the assay buffer (3 mg/10 mL). Prepare daily.

CDNB substrate: Prepare a stock solution at 100 mM: 0.1 g 1-chloro-2-4-dinitrobenzene in 5 mL ethanol, and dilute to 1 mM with the assay buffer.

7.5.2 Procedure

GST is a cytosolic enzyme and the assay is usually made in the post-mitochondrial fraction ($> 9000 \times g$) of the homogenate. The fractions could also be filtered on a 0.2 μm pore filter to remove microsomes and other membrane vesicles. There is no particular preference for the homogenization buffer, but it should contain a protease inhibitor buffered at pH 7.4 with Tris, HEPES, or phosphate buffer. In a clear microplate or borosilicate tube, mix 50 μL of buffer or $9-15,000 \times g$ supernatant, 100 μL of reduced GSH reagent, and 100 μL of CDNB substrate. Measure the absorbance at 340 nm at 5 min intervals for up to 30−40 min in a microplate reader. For a standard cuvette spectrophotometer, determine the absorbance at 30 min after adding 0.5−1 mL of water (depending on the required volume for the instrument) using the above mixture from which GSH was replaced by the assay buffer as the blank and maintained on ice.

7.5.3 Data Calculation

The rate of conjugation is calculated as follows:

$$(A \text{ sample} - A \text{ blank or } t = 0) \times 250/50 \times 1/\text{min} \times 1/\text{mL of supernatant used} = \text{increased absorbance}/(\text{min} \times \text{mL}),$$

the rate of increase is then normalized against total protein concentration or tissue weight (g tissue/mL of homogenization buffer) to give:

Absorbance at 340 nm/min/mg proteins or tissue wet weight.

7.6 URIDINE DIPHOSPHATE-GLUCURONYL TRANSFERASE ACTIVITY

The enzyme uridine diphosphate-glucuronyl (UDPG) transferase is a microsomal enzyme that catalyzes the conjugation of UDPG to oxidized (hydroxylated) xenobiotics. This enzyme is also involved in the inactivation of steroids such as estrone and testosterone [14]. The principle of the assay is based on the conjugation of UDPG with nitrophenol, which can be determined by absorbance at 405 nm: UDPG + nitrophenol → UDP-nitrophenyl + glucuronide. Increased activity of this enzyme is indicative of enhanced UDP conjugation to xenobiotics and sex steroids.

7.6.1 Reagents

Assay buffer: Prepare 40 mM Tris-HCl at pH 7.5 with the following: dissolve 0.48 g of Tris base in 90 mL of water, adjust pH to 7.5 with HCl 1 M, and complete to 100 mL with bidistilled water.

MgCl$_2$: Prepare 5 mM working dilution: dissolve 9.6 mg of anhydrous MgCl$_2$ in 20 mL of assay buffer. Prepare daily.

UDGGA co-substrate: Prepare a 150 mM stock solution: 87 mg UDP-glucuronic acid in 1 mL of assay buffer.

p-Nitrophenol: Prepare an 8 mM working solution: 22 mg in 20 mL of assay buffer. Triton X-100 activation: prepare 20% dilution in bidistilled water.

7.6.2 Procedure

Since the UDPG transferase is located in microsomes, the assay should be performed in the post-mitochondrial fraction or precipitated microsomes with the calcium aggregation method described previously or by ultracentrifugation. The reaction mixture contains 50 µL of either post-mitochondrial fraction or microsomes or blank (homogenization buffer), 20 µL of UDP-glucuronate (not in the blank), 20 µL MgCl$_2$, and 90 µL of assay buffer. The reaction is initiated with the addition of 20 µL of p-nitrophenol. The post-mitochondrial fraction or microsomes are previously treated with Triton X-100 at a concentration of 1 µL of 20% Triton X-100/mg proteins in the samples, respectively. The reaction is allowed to proceed at 20−30°C for 40 min. Absorbance readings at 405 nm are taken at each 10 min.

7.6.3 Data Calculation

$$(A\ 30\ min - A0\ min\ or\ blank)/30\ min \times 200/50 \times 1/volume\ of\ sample\ (mL)$$
$$= Change\ in\ absorbance/min/mL$$

The enzyme activity is then normalized by protein content (mg/mL) of soft tissue weight (g/mL of added homogenization buffer) to give: change in absorbance/min/mg proteins.

7.7 SULFOTRANSFERASES

Sulfotransferases are a family of enzymes involved in the conjugation of sulfates (SO_3^{-2}) to an acceptor alcohol or amine groups of xenobiotics. The most common biological sulfo acceptor group is 3'-phosphoadenosine-5'-phosphate (PAP) leading to the formation of 3'-phosphoadenosine-5'-phosphosulfate (PAPS) where the latter is involved in the formation of cysteine in cells. A spectrophotometric assay for sulfotransferase is presented where p-nitrophenyl sulfate is the sulfate donor and PAP is the sulfate acceptor molecule forming PAPS and p-nitrophenyl anion, which can be measured at 405 nm.

7.7.1 Reagents

Assay buffer: Prepare 50 mM NaCl containing 25 MOPS buffer, pH 7.2, and 1 mM EDTA.

PAP reagent: Prepare a 2 mM stock solution (8.6 mg/10 mL) in the assay buffer.

p-Nitrophenyl sulfate reagent: Prepare a 2 mM (5.2 mg/10 mL) in the assay buffer.

7.7.2 Procedure

Sulfotransferase activity is usually determined in the post-mitochondrial fraction (supernatant after $>9000 \times g$ centrifugation). Tissues are homogenized at 10−20% tissue weight/homogenization buffer volume ratio with a Teflon pestle tissue grinder. The homogenization buffer should contain a protease inhibitor and the pH maintained at 7−7.4. In a clear microplate or borosilicate tube, pipette 50 μL of S10 (10,000 × g supernatant for 20 min at 4°C) or water (blank), 25 μL of PAP reagent, and 25 μL of p-nitrophenyl sulfate and complete to 250 μL (150 μL) with the assay buffer. The formation of p-nitrophenyl anion was measured over 30 min at 405 nm at 10 min intervals if a microplate reader is available. For a standard spectrophotometer, water is added to 50 μL of the reaction mixture to accommodate the cuvette volume (500 μL or 1000 μL). In this case, the PAP reagent is omitted and replaced with the assay buffer as the blank samples.

7.7.3 Data Calculation

The rate of sulfate transfer is calculated as follows:

(A sample − Ablank or $t = 0$) × 250/50 × 1/30 min × 1/mL of supernatant used
$$= \text{increased absorbance at 400 nm/(min} \times \text{mL).}$$

The rate of increase is then normalized against total protein concentration (mg/mL) or tissue weight (g tissue/mL of homogenization buffer) to give the following:

Absorbance at 400 nm/min/mg proteins or tissue wet weight.

7.8 MULTIDRUG RESISTANCE GLYCOPROTEINS

Multidrug resistance proteins are determined with representative anionic and cationic organic molecules in cytoplasmic membrane vesicles. They are membrane bound P-glycoproteins that were responsible for increased resistance against environmental pollutants and tumor cells against chemotherapeutic agents [15,16]. These proteins represent one of the most formidable mechanisms of cell resistance to the toxic action of organic chemicals.

Fluorescein and rhodamine 123 are used as anionic and cationic substrates because they are cheap, safe to handle, and sensitively determined by fluorescence spectrometry. The process requires energy in the form of ATP, which can serve as another measurement endpoint (i.e., released phosphates from the hydrolysis of ATP). The principle of the assay is depicted in Figure 7.1. Upon homogenization, the cytoplasmic membranes are inverted inside out so that an efflux results in the accumulation of the dye inside the vesicles. After incubation of 30–60 min at room temperature, the incubation media is filtered by centrifugation where the dyes could be analyzed either in the eluate (remaining dye) or in the filtrate (retained dye in vesicles). The released inorganic phosphates could also be determined by the phosphomolybdate reagent (Chapter 10; vitellogenin assay by the alkali-labile phosphate method). The dye in the vesicles could be liberated by washing the filters with 20% DMSO.

7.8.1 Reagents

Homogenization buffer: 140 mM NaCl, 1 mM EDTA, 1 μg/mL apoprotinin, and 10 mM HEPES-NaOH, pH 7.4.

Assay buffer: Prepare 140 mM NaCl, 2 mM KCl, 1 mM CaCl$_2$, and 10 mM HEPES-NaOH, pH 7.4. Prepare before the assays.

Transport buffer: Add ATP to the assay buffer at a final concentration of 200 μM.

Dye reagent: Prepare either anionic or cationic dyes in separate tubes. For the anionic dye, prepare a 10 μM solution of dicarboxyfluorescein in the

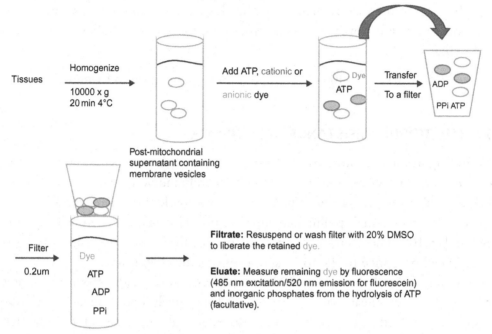

Figure 7.1 Principle of the efflux of xenobiotic assay.

assay buffer. For the cationic dye, prepare a 20 µM solution of rhodamine 123. A 1 mM stock solution could be prepared in the methanol by preparing 3.76 and 3.8 mg dicarboxyfluorescein or rhodamine 123 dissolved in 10 mL of methanol, respectively. Mix thoroughly and dilute further to obtain 50 and 100 µM in the assay buffer, respectively.

7.8.2 Procedure

The tissues (usually liver or gills) are weighted and homogenized at 20% density (tissue weight/buffer volume) in the described above media using four to six passes of a Teflon pestle tissue grinder. The homogenate is then centrifuged at 10,000 × g for 20 min at 2°C. The supernatant (S10) is carefully removed from the pellet and kept on ice. The transporter assay is then determined as follows. To 50 µL of S10, add 180 µL of transport buffer on ice. Remove and incubate for 5 min at 20−22°C. The reaction is initiated by the addition of 50 µL of the dye reagent and allowed to proceed for 30 min. The incubation mixture is then filtered on a 0.2 µm pore filter either through a syringe filter unit or an inserted filter for microcentrifuge tube as shown in Figure 7.1. The blank consists of mixing ice-cold S10 and transport buffer with the dye reagent and immediately filtered ("$T = 0$ min"). The fluorescence could be measured in either the filtrate (retained membrane vesicles) or the eluate (passed through the filter). For the eluate, the fluorescence is read for fluorescein (anionic transport) at 485 nm excitation and 520 nm emission and the same wavelengths for rhodamine 123 (cationic transport). A 1 µM or 2 µM standard solution for dicarboxyfluorescein or rhodamine 123 could be prepared for calibration. For the filtrate, the retained dye could be eluted from the vesicles by adding 20% DMSO in water. Fluorescence in the eluate is then measured as described earlier. Depending on the relative concentration of multi-drug resistant proteins in the sample, higher dye concentrations could be used or dilute the S10 with the assay buffer. Initially, the transport assay should be titrated against increasing concentration of the dye to ensure steady-state conditions. Indeed, the efflux pumps should be saturated with the dye to ensure constant activity. The rate of hydrolysis of ATP could also be followed by measuring the formation of inorganic phosphate (phosphate assay, alkali-labile phosphates; Chapter 10).

7.8.3 Data Calculation

For the eluate fraction the rate of fluorescein or rhodamine transport is calculated as follows:

(RFU 0 min − RFU 30 min) × (standard concentration/RFU standard− RFU blank) × 1/30 min × dilution factor to give nmol of dye uptake/min/mL.

The dilution factor is $250/50 = 5$. The activity is then normalized against total proteins in the S10 fraction (mg/mL) or by the tissue homogenate weight (g/mL) to give dye uptake nmol/min/mg proteins.

REFERENCES

[1] Omura T, Sato R. The carbon monoxide-binding pigment of liver microsomes. I. Evidence for its evidence for its hemoprotein nature. J Biol Chem 1964;239:2370−8.
[2] Manahan SE. Toxicological chemistry. Toxicological chemistry and biochemistry. 3rd ed. New York: Lewis publishers/CRC Press; 2003. p. 139−164.
[3] Thomas PE, Ryan D, Levin W. Improved staining procedure for the detection of the peroxidase activity of cytochrome P450 on sodium dodecyl sulfate polyacrylamide gels. Anal Biochem 1976;75:168−76.
[4] Ravindranath V, Anandatheerthavarada HK. Preparation of brain microsomes with cytochrome P450 activity using calcium aggregation method. Anal Biochem 1990;187:310−3.

[5] Cinti DL, Moldeaus P, Schenkman JB. Kinetic parameters of drug-metabolizing enzymes in Ca^{2+}-sediment microsomes from rat liver. Biochem Pharm 1972;21:3249—356.

[6] Philpot RM, Bend JR. Benzopyrene hydroxylase activity in hepatic microsomal and solubilized systems containing rabbit or rat cytochrome P-448 or P-450. Life Sci 1975;16:985—97.

[7] Gelboin HV, Huberman E, Sachs L. Enzymatic hydroxylation of benzopyrene and its relationship to cytotoxicity. Proc Natl Acad Sci USA 1969;64:1188—94.

[8] Burke MD, Mayer RT. Ethoxyresorufin: direct fluorimetric assay for a microsomal O-dealkylation which is preferentially inducible by 3-methylcholanthrene. Drug Metab Dispos 1974;2:583—8.

[9] Gagné F, Blaise C, André C. Occurrence of pharmaceutical products in a municipal effluent and toxicity to rainbow trout (*Oncorhynchus mykiss*) hepatocytes. Ecotoxicol Environ Saf 2006;64:329—36.

[10] Stresser DM, Blanchard AP, Turner SD, Erve JC, Dandenau AA, Miller VP, et al. Substrate-dependent modulation of CYP3A4 catalytic activity: analysis of 27 test compounds with four fluorometric substrates. Drug Metab Dispos 2000;28:1440—8.

[11] Gowland BTG, McIntosh AD, Davies IM, Moffat CF, Webster L. Implications from a field study regarding the relationship between polycyclic aromatic hydrocarbons and glutathione S-transferase activity in mussels. Marine Environ Res 2002;54:231—5.

[12] Summer KH, Wiebel FJ. Glutathione and glutathione S-transferase activities of mammalian cells in culture. Toxicol Lett 1981;9:409—13.

[13] Boryslawskyj M, Garrood AC, Pearson JT, Woodhead D. Elevation of glutathione-S-transferase activity as a stress response to organochlorine compounds, in the freshwater mussel, *Sphaerium corneum*. Marine Environ Res 1988;24:101—4.

[14] Rao GS, Haueter G, Rao ML, Breuer H. An improved assay for steroid glucuronyltransferase in rat liver microsomes. Anal Biochem 1976;74:35—40.

[15] Kurelec B. A new type of hazardous chemical: the chemosensitizers of multixenobiotic resistance. Environ Health Perspect 1997;105:855—60.

[16] Allen JD, Brinkhuis RF, Wijnholds J, Schinkel AH. The mouse Bcrp1/Mxr/Abcp gene: amplification and overexpression in cell lines selected for resistance to topotecan, mitoxantrone, or doxorubicin. Cancer Res 1999;59:4237—41.

Cellular Energy Allocation

François Gagné

Chapter Outline

The balance between energy expenses and reserves was shown to be a critical pathway of toxicity for many xenobiotics. Cellular energy is required to detoxify and repair xenobiotic-induced damage at the expense of energy reserves destined for growth, survival, and reproduction. This concept forms the basis of the dynamic energy budget of toxicants (DEBTox) model [1]. Toxic compounds are taken up by organisms where they affect metabolic processes involved in energy production, stores, and metabolism. These processes are fundamental toward an organism's health, survival, growth, and reproduction. Indeed, when organisms spend more and more energy to metabolize, defend against and eliminate xenobiotics, less and less energy is available for survival, growth, and reproduction. It was shown that exposure to environmental contaminants such as cadmium, tributyltin chloride, and linear alkylbenzene sulfonate leads to increased energy expenses with reduced energy stores in the form of lipids, sugars, and proteins in daphnids [2]. This was, in turn, predictive of reduced population growth. In another study with zebra mussels, the cellular energy allocation concept was examined along a pollution gradient in the field [3]. The study shows that reduced cellular energy allocation to stores was predictive of impacts at the population level (low recruitment) in the field.

In the context of climate change, higher fluctuations in temperature are expected to occur and the effects of pollution in the interaction of temperature with energy expenses and reserves should be examined more closely. Exposure of isolated mitochondria preparations to pharmaceutical products and municipal effluents led to increased temperature-dependent mitochondrial electron transport activity in *Elliptio complanata* mussels [4]. The increase in electron transport activity in mitochondria from a temperature of 4 to 20°C was significantly higher when pharmaceuticals and municipal effluents were present in the incubation media, which suggests more energy is spent per unit of temperature change. Furthermore, when mussels were exposed to municipal effluents (terminal aeration lagoon) increased mitochondrial electron transport activity was observed showing increased sensitivity to temperature changes in organisms exposed to municipal wastewaters. The methods proposed in this chapter to measure mitochondria respiration rates (electron transport) and energy levels (lipids, sugars, and proteins) could be practiced with virtually any organism from which mitochondria could be isolated.

8.1 ENERGY RESERVES

8.1.1 Total Carbohydrates

Total carbohydrates are determined by the anthrone reaction, which forms a proportionally strong green color in the presence of sugars [5]. It is a general, nonspecific, cheap, and quick assay for carbohydrates.

8.1.1.1 Reagents

Sucrose standard solution: Dissolve 10 mg sucrose in 10 mL SQ water (1 mg/mL). Dilute 1/20 in water to obtain 50 μg/mL (working solution).

Anthrone reagent: Work in fume hood. Dissolve 50 mg anthrone in 50 mL of concentrated sulfuric acid (H_2SO_4). Discard if coloration becomes darkish brown. Store securely in the dark.

8.1.1.2 Procedure

The assay is performed with tissue homogenates with no special treatment. In a clear microplate or test tube, add 10 μL of biological sample, blank, and working sucrose standard (50 μg/mL); 45 μL of SQ water; and 150 μL of the anthrone reagent. Heat at 80−85°C in a water bath for 15 min, cool at room temperature for 5 min, and blot dry the bottom of the microplate. Determine the absorbance at 620 nm. Dispose of the content carefully following local safety recommendation for acid waste.

8.1.1.3 Data Calculation

The amount of sugars in the sample is calculated as follows:

$$(A620 \text{ sample} - A620 \text{ blank}) \times (2.5 \text{ μg/mL Standard}/(A620 \text{ standard} - A620 \text{ blank})) \times 200/10 = \text{μg/mL total sugars or sucrose equivalents.}$$

The concentration is then normalized against total proteins (mg/mL) of the homogenate or tissue homogenate weight g/mL: μg sugars/mg proteins or μg sugars/g tissues. The added standard could vary or a complete standard curve could also be constructed between 1 and 20 μg/mL if the absorbance values of the sample deviate strongly ($> \pm 2 \times$ difference in absorbance between the sample and the standard value) from the absorbance values of the standard.

8.1.2 Lipids

Lipids are considered the most caloric form of energy and amenable to long-term storage in organisms where lipids could be readily mobilized upon demand, for example, during prolonged starving or active gametogenesis. The levels of total lipids were determined using the spectrophotometric assay using the sulfo-phospho-vanillin method [6]. This method is more sensitive and reliable than the gravimetric analysis for total lipids of chloroform or dichloromethane extracts of tissues. It is also easily downscaled to micro-plates to permit high-throughput analysis if an absorbance microplate reader is available.

8.1.2.1 Reagents

Concentrated sulfuric acid: H_2SO_4, keep in fume hood and wear safety glasses, gloves, and lab coat.

Phospho-vanillin reagent: First dissolve 0.06 g vanillin in 10 mL SQ water. Take 3.5 mL and carefully transfer to 60 mL phosphoric acid (under fume hood and wearing protective gear) and 5 mL of water. Always pour in the direction of concentrated acid to the water to eliminate strong exothermic reactions (explosions).

Lipid standard: Prepare 23.3 μL of Triton X-100 (density of 1.07 g/mL) in 200 mL of SQ water to obtain 12.5 mg/mL. Olive or canola oil could also be prepared in ethanol, but Triton X-100 is more soluble and easy to manipulate.

8.1.2.2 Procedure

All steps should be prepared in a fume hood wearing protective eyewear and gloves. In a clear microplate or borosilicate tube, mix 20 μL of blank, standard, or sample; 30 μL of concentrated sulfuric acid; and 150 μL of phospho-vanillin reagent. Incubate 10 min at 75–80°C, cool at room temperature for 5 min, and blot dry the bottom of the microplate. Measure the absorbance at 540 nm. A pinkish color will appear as the lipid content increases.

8.1.2.3 Data Calculation

The amount of lipids is determined as follows:

$$(A540 \text{ sample} - A540 \text{ blank}) \times (1.25 \text{ mg/mL standard}/(A540 \text{ standard} - Ablank)) \times 200/10 = \text{mg/mL lipids in the sample.}$$

The lipid content is then normalized against tissue weight in the homogenate (mg/mL) or total protein content (mg/mL) to give mg lipids/g tissues or mg lipids/mg proteins. The added standard could be adjusted or a complete standard curve could also be constructed between 0.3 and 3 mg/mL if the absorbance values of the sample deviate strongly from the absorbance values of the standard: $> \pm 2 \times$ difference in absorbance between the sample and the standard value.

8.1.3 Analysis of Lipid Profiles by ^{1}H NMR

Proton-based nuclear magnetic resonance (^{1}H-NMR) spectroscopy is a powerful way to determine the levels of various metabolites in biological tissues. One advantage is that it could be used to determine specific metabolites or to fingerprint the metabolite expression pattern. However, this technology is not sensitive enough, so it requires drying/evaporation steps to concentrate the samples. The advent of relatively inexpensive cheap and miniature ^{1}H-NMR instrument will certainly promote this approach throughout laboratories worldwide. With the present methodology, the whole spectrum of nonpolar metabolites (lipidome) is analyzed for toxic fingerprint analysis. The tissues could be first lyophilized to remove water from the sample. The tissues are then extracted in methanol/$CHCl_3$ or dichloromethane to separate the nonpolar from the polar metabolites [7,8]. The chloroform phase is then dried under N_2 stream and resuspended in deuterated chloroform before NMR analysis. In this respect, NMR analysis does not require extensive sample preparation and is nondestructive.

The principle of proton-NMR is based on the observation that a spinning charge generates a magnetic field. In the presence of an external magnetic field (produced by the instrument magnet), the magnetic movement of the spinning proton singlet either aligns or opposes the magnetic field. The difference in energy between the two spin states increases as the strength of the external field increases, hence the use of high-energy magnets from 40 to 900 MHz. However, magnets >200 MHz require superconducting magnets and cryogenic freeze (liquid nitrogen), which becomes labor intensive and cumbersome. Recently, a cheap, benchtop high-resolution (50 ppb) NMR system was capable of handling a $\leq$20 μL sample volume without cryofreezing. This system is commercially available and reasonably priced (at a fraction of the price of NMR systems), which will increase the opportunity to run NMR analyses in ecotoxicology laboratories.

8.1.3.1 Reagents

Homogenization buffer. Prepare 1.15% KCl containing 1 mM sodium phosphate, pH 7.2.

Extraction solvent: Chloroform or dichloromethane in methanol at a volume ratio of 3:1. Prepare 100 mL in fume hood with gloves and safety glasses.

Tetramethylsilane (TMS) reference: Prepare a 1-10% stock solution in d6-DMSO or $CDCl_3$: 1-10 mL in 1 mL deuterated solvent. This is the standard reference for nonpolar metabolites.

TMSP: Prepare a 1-10% stock solution in d6-DMSO or deuterated water (D$_2$O): 1-10 mg in 1 mL d6-acetone or DMSO. This is the standard reference for the polar metabolites.

8.1.3.2 Procedure

Remove tissues and blot or lyophilize to remove water (optional). Disrupt tissues using a Polytron tissue grinder in the homogenization buffer at a density of 20% tissue weight/mL buffer. Add 2−3 volumes of extraction buffer and mix for 10 min. Centrifuge at 3000 × g to separate the phases. If the phases are well separated, add again 1 volume of the extraction solvent and centrifuge again. Collect the organic phase (at the bottom) and evaporate under nitrogen stream for 1−2 h at 40−45°C. The aqueous phase could be collected and concentrated by lyophilization. The polar fraction could be resuspended in 100 μL d6-DMSO or D$_2$O containing 1% TMSP. However, polar metabolite analysis is outside the scope of this chapter.

Resuspend the organic fraction in 100 μL deuterated chloroform (CDCl$_3$) containing 1% TMS. Inject the appropriate amount of sample to a miniature ^{1}H-NMR instrument (20 μL is required for picoSpin from Cole Parmer) and make 100−200 scans until the appearance of consistent NMR spectrum (as in Figure 8.1).

8.1.3.3 Data Analysis

In this experiment, the spectrum could be analyzed to seek out differences in the proton shift pattern of expression to determine particular changes between samples (e.g., control vs exposed group). The signal intensity (area under the peak) and the chemical shift (ppm) are two important signals in the NMR scan (Figure 8.1A). The relative amount of protons in specific regions relative to the TMS reference could also be determined. For lipid analysis, it is noteworthy to determine the area under the methyl CH3-/CH2 band (just after the TMS signal) and the presence of conjugated glycerol (phospholipids) at 4.3−4.5 and 5.3−5.5 ppm for H at C2 and C1/C3 of the glycerol molecule (Table 8.1).

8.1.4 Protein Content

The levels of total proteins could be determined by a number of assays varying in degrees of complexity. For example, the absorbance at 280 nm in diluted SDS (0.2%)-denatured protein samples is a simple way to determine total protein content, but it lacks sensitivity and is subject to background interference. The evaluation

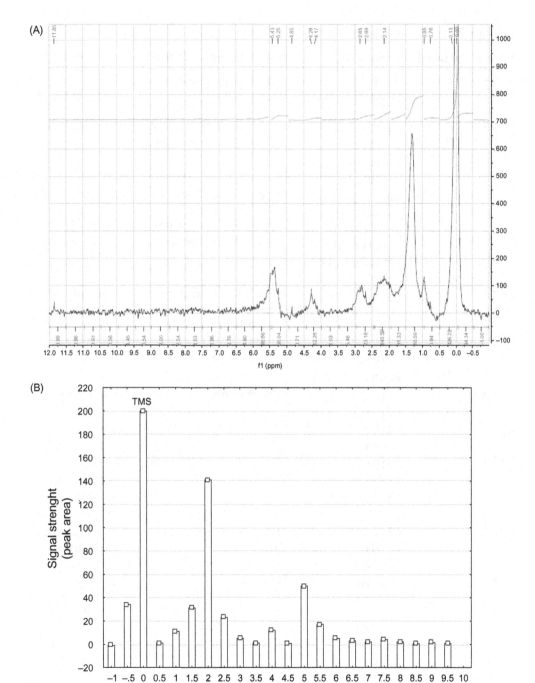

Figure 8.1 Typical ¹H-NMR spectrum of lipid extracts. A spectrum was taken using 45 MHz miniature high-resolution NMR system of a 20 μL lipid extract of flax seed (A). A number of 100 scans
were determined and the largest peak corresponds to the TMS reference (set at 0 ppm) and was
set at 0 ppm. The area under the signal at each 0.5 ppm interval was measured (B) using the data
analysis software of the instrument.

Table 8.1 Chemical Shift Data ^{1}H-NMR

Signal Range (ppm)	Compounds	Biochemical Interpretation	Comments
0–0.5	TMS	Highly shielded protons	
0.5–1	Saturated alkanes (CH3-R)	Saturation state (lipids)	Unsaturated lipids
1–1.5	CH2-R, CH-R, CH3-C-X x = halogen, OH, OR, Ar, N	Saturation state (linear and branched aliphatic chains)	Unsaturated lipids
1.5–2	Same above but with CH3-CR = CR2 or CH3-CR2-X x = halogen, OH, OR, Ar, N	Methyl from unsaturated lipids or presence of electroattractive groups (methyl deshielding),	Halogenation of unsaturated lipids (?)
2–2.5	CH3-Ar, CH3-C = O, nitrile R-C = C-H CH3-S or CH3-N R-OH	Methyl close to methoxy groups and presence of aromaticity	Methylation of alcohols (methoxy), thiols, and amines
2.5–3	CH3-O CH3-S CH3-N R-OH	Same as above	Same as above
3–3.5	CH3-O-R (ethers) R-OH HCR2-X	Oxidation of methyl groups	Secondary halogenated carbon Ether bonds
3.5–4	CH3-O-R (ethers) R-OH HCR2-X	Oxidation of methyl groups Glycerol (SN2)	Lipid degradation Inflammation
4–4.5	esters HC-O-C (=O)R R-NH2 NO2-CH Ph-O-CH R-OH Ar-OH	Fatty acids	Lipid breakdown: esterase or lipase activity

Continued

Table 8.1 Chemical Shift Data ^{1}H-NMR—(cont.)

Signal Range (ppm)	Compounds	Biochemical Interpretation	Comments
4.5−5	H2C = CR2	Terminal double bonds	Monounsaturated lipids
	Rn-C-OH	hydroxyl of glycerol	other: aromatic-OH could also emit at this range
5−5.5	PhOH	Hydroxyl aromatic (tyr)	
	Ar-OH s	Unsaturation of aliphatic chains	
	H-CR = CR2		
	Ar-OH	Glycerol (SN1 and SN3)	
5.5−6	R-CH = CH-R (conjugated)	Unsaturated lipids	Polyunsaturated compounds (lipids)
	Ar-OH		
	Amide RCONH		
6−6.5	H-C = C	Unsaturated lipids	
	R-CH = CH-R conjugated		
	Ar-OH deshielded		
	Amide RCONH shielded		
6.5−7.5	R-CH = CH-R conjugate	Unsaturated lipids	
9.5−10	R-CHO	Aldehydes	
	R-COOH	Carboxylic acids	
10−14	R-COOH	Carboxylic acids	

of total nitrogen from the Kjeldahl method in fractionated proteins is a definitive assay for proteins, but the method is less amenable to large-scale screening purposes. Total protein content could be determined by a simple, rapid, and inexpensive protein-dye binding assay [9]. The assay is amenable in microplates and the dye preparation could also be used for protein staining in gel electrophoresis.

8.1.4.1 Reagent

Coomassie blue reagent: The reagent could be prepared in-house by dissolving 50 mg of Coomassie blue G250 in 25 mL methanol. The methanol solution is then mixed with 50 mL of concentrated H_3PO_4 (prepare in fume hood and wear protective gear, gloves, and eyewear). The brownish mixture is then passed through a Whatman #2 filter paper (placed in a glass funnel over a beaker). The filtration step could take $1-2$ h to complete. Complete to 100 mL with bidistilled water. Store in dark brown bottle at 4°C.

NaOH solution: Prepare a 100 mM NaOH solution (100 mL: 0.4 g/100 mL distilled water).

Albumin standard: Prepare a 0.02% albumin solution in 100 mL of distilled water or in phosphate buffered saline (140 mM NaCl, 5 mM KH_2PO_4, 5 mM $NaHCO_3$, adjust pH to 7.4).

8.1.4.2 Procedure

In a clear microplate, prepare the following in triplicates. Ten microliters of test sample (homogenate or supernatants or water for the blank) and 10 µL of NaOH solution and wait for 15 min (optional step). Add 140 µL of water and 40 µL Coomassie blue reagent, mix well, and read absorbance at 595 nm after 5 min. The addition of NaOH serves to denature the proteins to favor a more homogenous binding of dye in the sample but increases the background (blank) absorbance. However, the assay works well without this step but could produce more variation in the absorbance values depending on the nature of the test sample.

For calibration purposes, an external calibration curve is prepared by the following steps:

Albumin (µg/mL)	Amount of Albumin Standard (µL)	NaOH (µL) Optional*	Water (µL)	Coomassie Reagent (µL)
0	0	10	150	40
5	5	10	145	40
10	10	10	140	40
20	20	10	130	40
25	25	10	125	40

*Optional step: replace with water if not used (160 µL of water instead of 150 µL).

8.1.4.3 Data Calculation

From the external standard measurements, a linear relationship is obtained between the albumin and the absorbance value at 595 nm. For example, the following linear equation could be obtained:

Absorbance at 595 nm $= 0.005 + 0.475$ (Protein concentration (µg/mL).

The protein concentration is then obtained by rearranging the equation to give the following:

Protein concentration $(µg/mL) = (A_{595} - 0.005)/0.475.$

The obtained protein concentration is then multiplied by the dilution factor of the assay 200/10 if the test sample was not pre-diluted in water. In many cases, the samples are diluted between 1/10 and 1/50 in water before. For example, if the test sample was diluted 1/20 in water and 10 μL of this dilution was used for the protein assessment, then the obtained protein concentration is multiplied by μg/mL proteins × (200/10) × 20 to obtain the final protein concentration of the sample.

8.1.5 Mitochondria Electron Transport Activity and Temperature Dependence

The assay is based on the relationship between mitochondria respiration rates (CO_2 production) and electron transport activity, which can be easily determined by an electron acceptor chromophore [10]. The assay is rapid and conveniently determined by standard spectrophotometry. It gives a measure of mitochondrial electron chain transport activity, which is a measure of energy expenses since glucose consumption is coupled to cellular respiration.

8.1.5.1 Reagents

Homogenization buffer: Prepare 20 mM Tris-HCl, pH 8, containing 100 mM NaCl (140 mM NaCl if the organism is a vertebrate), 0.1 mM EDTA, 0.1 mM dithiothreitol, and 1 mM KCl.

Assay buffer: Prepare 100 mM Tris–HCl, pH 8.5, containing 0.1 mM $MgSO_4$, 0.1% Triton X-100, and 5% polyvinylpyrrolidone.

NADH/NAPDH solution: Prepare reduced NADH and NADPH at 10 and 2 mM concentration, respectively, in the assay buffer. Prepare fresh and keep on ice (4°C).

Dye reagent: Prepare a 5 mM solution of p-iodonitrotetrazolium salt in the assay buffer.

8.1.5.2 Procedure

First, mitochondria should be isolated at sufficient density for detection. At least 500 mg tissues are minced with scissors and homogenized in 2.5 mL (20% tissue weight/volume density) ice-cold buffer using a Teflon pestle tissue grinder (4-6 passes). The homogenate is centrifuged at 1500 × g for 10 min at 2°C and the supernatant centrifuged at 9000 × g for 20 min at 2°C. The pellet is removed from the supernatant and resuspended in a small amount of the homogenization buffer. The mitochondrial fraction could be centrifuged again for washing purposes if a dense supernatant is obtained depending on the tissues and species. Determine the protein content of the mitochondrial suspension using the Coomassie blue dye binding assay as previously described. Adjust the protein concentration to 0.5−1 mg/mL with the homogenization buffer.

In a clear microplate or microcentrifuge tube, add 50−100 μg/mL of mitochondria (20 μL) with 80 μL of assay buffer and let stand for 5 min at room temperature. Add 20 μL of NADH/NADPH solution or water (blank) and 100 μL of the dye reagent. Read at 495 nm for each 10 min interval for up to 40 min at room temperature (20°C or 25°C exactly). For cuvette spectrophotometer, add buffer to accommodate the minimal cuvette requirements of the instrument and measure at 495 nm. The water blank serves as the control. For temperature dependence, repeat the assay at 4−6°C.

8.1.5.3 Data Calculation
8.1.5.3.1 Electron Transport Activity (at 25 and 4−6°C)
The rate of dye reduction is determined as follows:

$$(A\ 495\ 30\ min - A\ 495\ t_0) \times (220/20) \times 1/30\ min \times 1/$$
$$\text{added mitochondria volume (mL)} = \text{increase in absorbance}/(min \times mL),$$

the activity is then normalized against total proteins content of the mitochondria suspension (mg/mL) to yield:

Electron transport activity expressed as increase Absorbance/min × mg proteins.

8.1.5.3.2 Temperature-Dependent Electron Transport Activity
The metric is the difference between electron transport activity at 25 and 4−6°C against the variation in temperature:

$$(\text{Activity at } 25°C - \text{activity at } 6°C)/(25-6)°C = \text{Activity rate change}/°C,$$

which is an indicator of the susceptibility to temperature changes, i.e., to determine whether organisms spend more energy per temperature increment.

8.1.6 Cytochrome c Oxidase Activity

Cytochrome c oxidase (CCOX) is the last enzyme in the mitochondrial electron transport chain pigments. CCOX accepts an electron from cytochrome c proteins and transfers it to one oxygen molecule to form water. During that process the enzyme translocates protons (H^+) across the membrane and contributes to the transmembrane electrochemical gradient required for the synthesis of ATP (ATP synthase). The activity in this enzyme in a biological sample could be used to determine the number of active mitochondria or increase in electron transport activity in mitochondria (energy expense or respiration rate).

8.1.6.1 Reagents

Homogenization buffer: Prepare 250 mM sucrose with 10 mM Tris-HCl, pH 7.2, 1 mM EDTA, and 1 mM dithiothreitol. Adjust pH at 7.2 with HCl 1 N.

Assay buffer (5 ✕): Prepare 50 mM Tris-HCl and 600 mM KCl in 90 mL of water. Adjust pH to 7.2 with HCl (1 N) and complete to 100 mL with SQ water. Dilute 1/5 with water on the day of analysis.

Enzyme dilution buffer (2 ✕): Prepare 20 mM Tris-HCl and 500 mM sucrose in 90 mL of water. Adjust pH to 7.2 with HCl (1 N) and complete to 100 mL with SQ water. Dilute with 1 volume of SQ water at the day of analysis.

Reducer: Prepare a 0.1 M stock solution of dithiothreitol or β-mercaptoethanol (prepare in fume hood) or sodium dithionite in 10 mL of SQ water. Store at $-20°C$ in 1 mL aliquots.

Cytochrome c: Weight 27 mg in 10 mL of water and store in 1 mL aliquots at $-20°C$. At the day of analysis, mix 5 μL of reducer with 1 mL of cytochrome *c* solution and incubate for 15 min at room temperature. The reduction of cytochrome *c* is measured at 550 nm with a minimum at 565 nm. The absorbance ratio of 550/565 of 8 to 10 should be obtained. If the ratio remains <8 then add another 5 μL of the reducer.

8.1.6.2 Procedure

Tissues are homogenized in isotonic sucrose using a Teflon pestle tissue grinder (4−6 passes at 4°C). The homogenate is centrifuged at $1500 \times g$ for 10 min and supernatant collected. The supernatant is then centrifuged at $9000 \times g$ for 20 min and the supernatant is removed from the pellet containing crude mitochondria. The pellet is resuspended with the homogenization buffer and stored on ice or at $-80°C$ until analysis. The protein levels should be determined using the protein assay procedure provided in this chapter.

A sample volume of 10 μL is mixed with 180 μL of assay buffer and the reaction is started by the addition of 10 μL of reduced cytochrome c. The decrease in absorbance at 550 nm is taken at each 10 min interval for up to an hour at 25°C [11]. The blank contains the enzyme dilution buffer instead of the mitochondrial fraction of the homogenate.

8.1.6.3 Data Analysis

The enzyme activity is determined as follows:

$$(A550 \text{ ''0'' min} - A550 \text{ 30 min}/30 \text{ min}) \times 1/0.01 \text{ mL}$$
$$= \text{Decrease Absorbance at 550 nm/min} \times \text{mL}.$$

The activity is then normalized against total proteins in the mitochondrial fraction (mg/mL) to give decreased absorbance at 550 nm/min/mg proteins.

REFERENCES

[1] Jager T, Heugens EHW, Kooijman SALM. Making sense of ecotoxicological test results: towards application of process-based models. Ecotoxicology 2006;15:305–14.

[2] De Coen W, Janssen CR. The missing biomarker link: relationships between effects on the cellular energy allocation biomarker of toxicant- stressed *Daphnia magna* and corresponding population characteristics. Environ Toxicol Chem 2003;22:1632–41.

[3] Smolders R, Bervoets L, De Coen W, Blust R. Cellular energy allocation in zebra mussels exposed along a pollution gradient: linking cellular effects to higher levels of biological organization. Environ Pollut 2004;129:99–112.

[4] Gagné F, Blaise C, André C, Salazar M. Effects of pharmaceutical products and municipal wastewaters on temperature-dependent mitochondrial electron transport activity in *Elliptio complanata* mussels. Comp Biochem Physiol C Toxicol Pharmacol 2006;143:388–93.

[5] Jermyn MA. Increasing the sensitivity of the anthrone method for carbohydrate. Anal Biochem 1975;68:332–5.

[6] Frings CS, Fendley TW, Dunn RT, Queen CA. Improved determination of total serum lipids by the sulfo-phospho-vanillin reaction. Clin Chem 1972;18:673–4.

[7] Viant MR. Environmental metabolomics using ^{1}H-NMR spectroscopy. In: Martin CC, editor. Methods in molecular biology: environmental genomics. Totowa, NJ: Humana Press Inc; 2008. p. 137–52.

[8] Wu H, Southam AD, Hines A, Viant MR. High-throughput tissue extraction protocol for NMR and MS-based metabolomics. Anal Biochem 2008;372:204–12.

[9] Bradford MM. A rapid and sensitive method for the quantitation of microgram quantities of protein utilizing the principle of protein-dye binding. Anal Biochem 1976;72:248–54.

[10] King F, Packard TT. Respiration and the activity of the respiratory electron transport system in marine zooplankton. Limnol Oceanog 1975;20:849–54.

[11] Sakai Y, Tanaka A, Ikai I, Yamaoka Y, Ozawa K, Orii Y. Measurement of cytochrome c oxidase activity in human liver specimens obtained by needle biopsy. Clin Chem Acta 1988;176:343–6.

CHAPTER *9*

Neuroendocrine Disruption

François Gagné

Chapter Outline

Biochemical Ecotoxicology
DOI: http://dx.doi.org/10.1016/B978-0-12-411604-7.00009-X

The presence of chemicals that can disrupt neuroendocrine signaling systems is an area of intense research in ecotoxicology. Disruption in estrogen signaling by estrogen-mimicking compounds was and still is the object of many investigations, but not only hormones are affected, neurotransmitters are as well [1,2]. In oviparous organisms, stimulation of the estrogen receptor (estradiol-17β) leads to the synthesis of vitellogenin, an egg yolk precursor destined as an energy source for the developing embryo. Prolonged exposures to these estrogen mimics could lead to altered sexual development such as intersex and male feminization. Recent evidence suggests that serotonergic and perhaps opiate-like compounds could also be found in municipal wastewaters, which constitute other types of neuroendocrine disrupters that could impede the reproductive and behavioral status in wildlife [2]. This chapter will present methods for examining the neuroendocrine system in fish and bivalves at various aspects: estradiol-17β binding, vitellogenin, serotonin and dopamine metabolism and signaling, glutamate and γ-aminobutyrate metabolism and acetylcholinesterase.

9.1 VITELLOGENIN EVALUATION AS A MARKER OF ESTROGENICITY

Vitellogenin (Vtg) is synthesized in the liver in vertebrates and in the gonad tissues in invertebrates (mussels, clams, and gastropods) and is under the control of estradiol-17β and other neuropeptidic precursors from the nervous system. Indirect and direct assays for Vtg are presented in this chapter. The indirect assay is based on the levels of alkali-labile phosphates (ALP) in high molecular weight proteins, which are fractionated by either acetone or t-butyl methyl ether [3]. Indeed, ALPs could be used as a surrogate for Vtg levels in fish, mussels, and clams but not in gastropods in most cases. There are many variations of the ALP method for Vtg assessment, which brings some degree of sensitivity and offers less interference such as turbidity [4,5]. Direct assays for Vtg could be performed by quantitative polymerase chain reaction or by commercially available enzyme-based immunoassays if available for a given species. Vtg is highly species specific so that a salmonid Vtg immunoassay usually does not work with other fish species, for example.

9.1.1 Reagents

Phosphate standard: Dissolve 1.74 g of monobasic potassium phosphate in 100 mL of NaOH 1 M to give 0.5 g phosphate/100 mL (5 mg of phosphate/mL). Dilute this stock solution 1/100 to obtain a final concentration of 50 ug Phosphate/mL.

Molybdate reagent: Dissolve 1 g ammonium molybdate (Sigma A7302) in 100 mL of 0.9 M sulfuric acid (H_2SO_4). Make sulfuric acid at this concentration by mixing 5 mL of concentrated sulfuric acid (in a fume hood and wear protective gloves, eye glasses, and lab coat) in 100 mL of SQ water.

Ascorbate reagent: 1 g of L-ascorbic acid (Sigma A7631) with 100 mL H_2O.

NaOH 1 M: Dissolve 4 g NaOH in 100 mL of SQ water.

9.1.2 Procedure

Tissues are homogenized in ice-cold homogenization buffer containing protease inhibitor, reductive agents (dithiothreitol or β-mercaptoethanol), and calcium/magnesium chelator (EDTA); for example, 50 mM Tris-acetate buffer, pH 8, containing 100 mM NaCl, 0.1 mM dithiothreitol, and 0.1 mM EDTA. Apoprotinin could be added as a protease inhibitor at 1−10 μg/mL. The homogenate is then centrifuged at 12,000 × g for 20 min at 2°C and supernatant collected for analysis.

Acetone is added to 200 μL of the supernatant to obtain a final concentration of 35% and placed on ice for 10-15 min. The mixture is then centrifuged at 10,000 × g for 5 min at 4°C. The supernatant is removed and the pellet washed in 500 μL of 50% acetone to remove any trace of lipids and free phosphates. The pellet is then dried (air) and resuspended with 100 μL of NaOH 1 M and incubated for 60 min at 37°C.

The released inorganic phosphates are then determined by a spectrophotometric assay using the phosphomolybdic acid complex reagent [6]. In a clear microplate, 20 μL of the blank (NaOH alone), standard (20 μL of phosphate standard 5 µg/mL phosphate), and sample (pellet resuspended in NaOH) is added in duplicate to 125 µL of water. Add quickly 5 µL of trichloroacetic acid (6.1 N), 25 µL of molybdate reagent, and 25 µL of ascorbate, and mix thoroughly for 5 min in a microplate mixer if available or by repeated pipetting if not. Determine the absorbance of the phosphomolybdate complex at 815 nm. The absorbance at 444 nm is also measured for turbidity correction if present.

9.1.3 Data Calculation

First, the absorbance at 815 nm is corrected for turbidity by the following equation:

$$A815 \text{ corrected} = (1.045 \times A815) - (0.043 \times A444).$$

The phosphate concentration is determined by the following:

[A815 sample − A815 blank]
  × Standard phosphate concentration/[A815 standard − A815 blank]
  × (sample volume µL/total assay volume µL) × (1 mL /S12 volume used)
  = Y µg/mL.

In the present example, the volumes are, respectively, $200/20 \times 1000/200 = 50$. This concentration is then normalized against total proteins (mg/mL) or homogenate tissue weight in g/mL:

Y µg/mL/g tissue weight /mL or mg/mL protein
  $= Z$ µg phosphate/g or Z µg phosphates/mg proteins).

9.1.4 Precautions

The generic assay for Vtg using the ALP is based on the principle that when the highly phosphorylated Vtg is produced in tissues at high concentration. The levels of ALPs in high molecular weight lipophosphoproteins increase above the background levels of ALP in tissues. Given that egg yolk proteins could reach more than 50% of total proteins in the gonad, the increase in ALP from high molecular weight proteins is indicative of Vtg. In addition, Vtg proteins generally have a higher proportion of phosphorylation sites than the average levels in tissues in the attempt to mitigate background signals, but this depends on the species under examination. Indeed, it was estimated that the mean value of serine and threonine in a protein pool in eukaryotes is in the range of 10.9% of total amino acid content [7]. For example, Vtg from *Mytilus edulis* has about 14.8% of potential phosphorylation sites (serine and threonine) that are compatible with the ALP methodology. Conversely, in gastropods, the number of phosphorylation sites is close to the background mean value and the ALP method cannot be used [8].

The ALP technique is a convenient, cheap, and rapid means to screen or explore the presence of high molecular weight phosphorylated lipophospho-proteins in various organisms. However, since this method is indirect, we recommend running gel electrophoresis with either silver or Coomassie blue staining on the acetone-precipitated material to detect the presence of specific female proteins in gonad tissues to validate the assay. Phosphate staining using various dyes (ammonium molybdate/methyl green, "Stains all," Pro-Q Diamond stain) could also be used. The following precautions should be considered when applying the indirect ALP assay in organisms:

1. Female-containing oocytes should have higher ALP values than background, which could be either nongravid females or males. This could be checked by running a gel electrophoresis in parallel at first.

2. Because the baseline ALP values are background (from the mean phosphate levels of other proteins during acetone fractionation), we cannot interpret this as Vtg obviously; it is the rise (change) in ALP from fractionated high molecular weight lipophosphoproteins that is related to Vtg and not the baseline values.

3. The ALP data should be compared between males and females that are at the same gametogenesis stage (early, mid, late, and resting stages). ALP levels in females at early-mid stages of gametogenesis should be higher than males. In some cases, if protein contents in females are higher in gonad than in males at the resting or early stages, then the ALP levels in females could be lower than males.

4. It is generally recognized that organisms under active gametogenesis at the late and spawning stage are less sensitive to exposure to endocrine disrupters because their physiology is driven by reproduction. For the detection of endocrine disrupters, ALP levels are usually measured in organisms at the resting or early stage of gametogenesis and preferably in males. For example, it was reported that wild mussel populations collected downstream to

municipal wastewater dispersion plumes showed increased proportion in females whereas males had increased ALP levels and showed the female-specific high molecular weight protein band after gel electrophoresis [9].

5. In gravid females or at mid to late stages of gametogenesis, the ALP method could suffer from over normalization problems when normalizing by total protein content. If egg yolk proteins represent more than 1% of total protein content in the sample, normalizing with total proteins could over normalize the ALP levels. We recommend correcting against tissue weight or using the residual or stepwise addition methods explained in Chapter 2.

9.2 ESTRADIOL-17β OR STEROID BINDING SITES

The occurrence of estradiol-17β binding sites could be determined by the binding potential of estradiol-17β from cytosolic proteins in a given tissue extract. To measure the binding to cytosolic proteins, estradiol-17β linked to fluorescently labeled albumin (E2-A-FITC) was used. E2-A-FITC linked to receptors will produce a protein complex that can be removed from unbound E2-A-FITC by ultrafiltration through a 100 kDa pore membrane. Indeed, E2 receptor has a molecular weight of ∼70 kDa and albumin a molecular weight of 60 kDa equaling a protein complex of 130 kDa, which can be separated by a 100 kDa ultrafiltration device (Figure 9.1). Care was taken to limit unspecific protein interaction and free estradiol-17β was found to compete effectively with E2-A-FITC binding [10]. In *Elliptio complanata* gonad, a dissociation constant of 0.4 nM for estradiol was determined by this methodology, which was in the same range of amphibian receptors [10].

9.2.1 Reagents

Receptor isolation buffer: 10 mM HEPES, pH 7.4, containing 50 mM NaCl, 10 mM ammonium molybdate, 1 mM EDTA, and 5 mM dithiothreitol.

Assay buffer: PBS containing 0.1% Tween 20, 0.05% bovine serum albumin (BSA), and 5 mM dithiothreitol. Dissolve 0.1 g of Tween 20, 0.1 g BSA, and 77 mg of dithiothreitol in 100 mL of PBS solution. PBS is composed of 140 mM NaCl, 5 mM KH_2PO_4, 1 mM $NaHCO_3$, and 10 mM HEPES-NaOH buffer, pH 7.4. Filter on 0.2 μm membrane filter. Prepare fresh.

E2-A-FITC reagent: Dissolve 1 mg of E2-A-FITC (E2 linked to albumin fluorescein isothiocyanate) in 10 mL of assay buffer. Store in the dark for

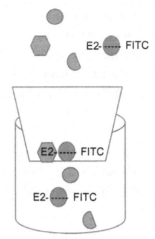

Figure 9.1 Schematic principles of estradiol-17β binding sites in tissues. Free E2-A-FITC passes through the filter while E2-A-FITC bound to protein receptors is retained on the 100 kDa membrane filter.

1 week at 4°C. Dilute to 100 nmol estradiol/mL with the assay buffer. A reducer such as dithiothreitol is required since albumin tends to oxidize and form aggregates, which increase the background binding of E2-A-FITC to the membrane filter.

9.2.2 Procedure

Tissue homogenates are prepared in the receptor isolation buffer using a Teflon pestle tissue grinder (five passes on ice). The homogenate is then centrifuged at $12,000 \times g$ for 20 min at 2°C. The supernatant or S12 fraction is then recuperated, kept on ice, and filtered on a 0.2 μm membrane filter to remove membrane vesicles. The filtrated S12 fraction is then diluted 1/5 in the assay solution. An equal volume (150 μL) is mixed with 1 volume of E2-A-FITC reagent and incubated for 30–60 min at room temperature. The mixture is then centrifuged on 100 kDa ultrafiltration device (400 μL ultrafiltration insert unit for microcentrifuged tubes; Sigma Chemical Company) at $10,000 \times g$ at 4°C for 15–30 min. E2-A-FITC bound to receptors or other high-affinity proteins (i.e., with molecular weight of 40 kDa or more) will be retained on the ultrafilter membrane while unbound or free E2-A-FITC will pass through the membrane. Fluorescence associated with the membrane or in the eluate is measured using a microplate reader at 485 nm excitation and 512 nm emission. For measurement in the eluate, the inverse relationship of fluorescein is a measure of E2 binding to cytosolic proteins. Blanks contain only

E2-A-FITC with no cytosolic proteins and positive control of unlabeled estradiol-17β could be used at concentrations of 5–100 nmol/mL to check for the release of unbound E2-A-FITC. For measurement on the ultrafiltration membrane, the material is recuperated by adding 100 μL of PBS on the membrane and resuspended by repeated pipetting or purged with 20 % DMSO in water.

9.2.3 Data Calculation

The concentration of fluorescein in the eluate (unbound E2-A-FITC) is determined as follows: fluorescein found in the eluate (free form of E2) and corrected with the levels of total proteins. Data are expressed as nmol/mL fluorescein-1/mg protein.

9.3 FLUOROMETRIC ASSAY FOR CATECHOLAMINES AND INDOLAMINES

Catecholamines (dopamine and adrenaline) and indolamines (serotonin) can be determined by fluorescence spectroscopy. This assay is a rapid and cost-effective way to determine the levels of the neurotransmitters described above provided a fluorometer is available. This method is based on the fractionation of the amines and selective fluorescence determinations for serotonin [11] and catecholamines [12]. Serotonin is determined using the o-phthaldehyde reagent, which reacts with secondary amines in the presence of excess thiols (R-SH).

9.3.1 Reagents

Homogenization buffer: Tissues are homogenized in any isotonic buffer such as 250 mM sucrose containing 10 mM HEPES-NaOH, pH 7.4, containing 1 mM EDTA and 1 mM dithiothreitol as reducing agents.

Reduced glutathione (GSH$_R$): Prepare a reduced glutathione stock solution at 100 mM in the presence of 1 M HCl. Stable in the dark at 4°C for one week. Dilute 1/100 on the day of analysis by mixing 20 μL stock in 2 mL of water.

n-butanol solvent: Reagent grade n-butanol.

OPT reagent: Dissolve 10 mg of o-phtaldialdehyde in 0.1 M HCl. Keep in the dark and stable for one week at 4°C.

Lugol's solution: Dissolve in a fume hood 2.5 g of iodine and 5 g of potassium iodide in 100 mL distilled water. This solution is also commercially available at different formulations.

Standards: Prepare separately a 5 μmol/mL (5 mM) solution of serotonin, dopamine, 5-hydroxyindolacetic acid (5-HIAA), and adrenaline in 0.1 M HCl. These biogenic amines, with the exception of 5-HIAA, are subjected to oxidation, which is prevented in acid conditions. The addition of 1 mM reduced glutathione or other reducer could further prevent oxidation. Stable for 1 week in the dark at 4°C.

NaHCO₃ solution: Prepare a 100 mM NaHCO$_3$ concentration in water, adjust the pH at 9 with NaOH 1 M.

9.3.2 Biogenic Amine Extraction

In duplicate microcentrifuged tubes, add 100 μL of GSH$_R$ to 250 μL of homogenate. Mix tubes by inversion and stand for 5 min. Add 300 μL of n-butanol and mix (vortex) and centrifuge at 10,000 × g for 20 s. Remove the supernatant and discard the pellet. Add 1 volume of hexane (600 μL) and centrifuge again. The aqueous phase is at the bottom and the organic phase at the upper layer. The metabolite of serotonin (5-HIAA) is found in the organic phase (keep aside).

A facultative preconcentration step could be done if the biogenic amines are too diluted in the homogenate. The procedure consists of mixing 200−300 μL of the aqueous phase to 1 volume of water and 50 mg of activated alumina slurry (washed in bidistilled water) in a microcentrifuge tube. The alumina is pelleted by brief centrifugation at 5000 × g for 30 s and resuspended in water. The sample is centrifuged again, the water is discarded, and the biogenic amines are eluted by resuspending the pellet in 50−100 μL of 100 mM Tris-acetate, pH 8. The alumina is removed by a final brief centrifugation step.

9.3.3 Determination of Indolamines (Serotonin)

In a dark microplate for fluorescence readings, pipette 25−100 μL of the aqueous phase (lower phase), and add 100 μL of SQ water and 40 μL of the OPT reagent. Heat the microplate at 70°C in a water bath for 15 min. Cool at room temperature and determine fluorescence at 360 nm excitation and 460 nm for emission. The blank contains only the homogenization buffer and reduced GSH, and 0.5 mM serotonin standard is added in another well in duplicate for calibration.

9.3.4 Determination of 5-HIAA

In a microcentrifuge tube, mix 200 μL of the upper hexane phase with 100 μL of NaHCO$_3$. Mix by inversion and centrifuge briefly to separate the phases if necessary. Collect 75—85 μL of the aqueous phase at the bottom of the tube. Perform assay as described previously for serotonin. A blank and 0.5 mM standard solution of 5-HIAA is prepared by replacing the homogenate sample with NaHCO$_3$.

9.3.5 Determination of Adrenaline and Dopamine

In a microcentrifuge tube, remove 100 μL of the aqueous phase and add 390 μL of SQ water and 10 μL of alkaline iodine/iodide solution (Lugol's solution). Prepare this in a fume hood and wear protective eyewear and gloves. Incubate for 5 min at 75°C and transfer 200 μL to a fluorescence microplate reader or spectrometer and read fluorescence at 395 nm excitation and 475 nm for emission (adrenaline). Prepare blanks and adrenaline standard in separate wells (50—150 nmol/mL). Continue heating at 75°C for 60 min and remove 200 μL to a fluorescence microplate/spectrometer and read at 325 nm excitation and 380 nm for emission (dopamine). Prepare blanks (homogenization buffer only) and dopamine standard (50—150 nmol/mL) in separate wells.

9.3.6 Data Calculation

The data are obtained using the serotonin protocol as an example:

$$(\text{Fluorescence units (FU) sample} - \text{FU blank})$$
$$\times\, [\text{Serotonin standard concentration (500 nmol/mL)}$$
$$/(\text{FUstandard} - \text{FUblank})] = \text{nmol/mL serotonin}.$$

This concentration is then corrected for dilution factors (240/100) × (1000/200 μL homogenate volume) and normalized against total proteins of homogenate or wet/dry weight to yield nmol serotonin/mg proteins or g weight.

9.4 COMPETITIVE IMMUNOASSAYS FOR DOPAMINE AND SEROTONIN

A highly sensitive, specific, and rapid assay for dopamine and serotonin in biological samples is presented. The assay is based on the competitive binding of dopamine monoclonal antibody to dopamine bound at the bottom of the well and dopamine present in the test sample [13]. The competitive assay is based on the following principle: the more concentrate the dopamine in the test sample, the less antibody will be bound to dopamine adsorbed at the bottom of the well. The amount of antibody attached at the bottom of the well is detected by a secondary antibody-linked horseradish peroxidase. To further increase the sensitivity of the assay, peroxidase activity is detected using a highly sensitive chemiluminescent substrate (luminol). This assay therefore requires a luminescence microplate reader. Fluorescence- and absorbance-based substrates could be used instead but with lower sensitivity. The assay is presented for dopamine and could be adapted to serotonin (or any other neurotransmitter or hormone) by just changing the primary antibody and standards for serotonin as described in the following tables.

9.4.1 Reagents and Solutions

Items

Phosphate buffered saline (PBS)	140 mM NaCl, 5 mM KH_2PO_4, 1 mM $NaHCO_3$, pH 7.4
Coating buffer	50 mM Tris-HCl, pH 8.5
Washing buffer	PBS 1 ×
Blocking buffer	PBS with 1% fat-free milk powder
Dilution buffer	PBS with 0.5% fat-free milk powder
Dopamine—BSA (0.5 μg/well) Serotonin—BSA (0.5 μg/well)	5 μg/ml stock solution in water; stable for 1 week.
Primary antibody	Rabbit polyclonal to dopamine (IgG; raised against dopamine-glutaraldehyde-BSA) or rabbit polyclonal to serotonin (IgG) Dilute 1/5000 with dilution buffer
Secondary antibody	Anti-rabbit IgG: horseradish peroxidase conjugate Dilute 1/10000 with dilution buffer
Stock standard Dopamine 20 mM Serotonin 20 mM	Prepare in PBS (unstable, keep in the dark and at 4°C)
Peroxidase substrate	BM Chemiluminescence ELISA substrate

9.4.2 Procedure

First, prepare serial dilutions of dopamine using the dilution buffer as described in the following table.

	Dopamine (μM)	Volume (μl) Dopamine 20 mM	Volume (μl) Dilution Buffer
S1	4000	300	1200
Serial dilutions			
S2	2000	750 μl S1	750
S3	1000	750 μl S2	750
S4	500	750 μl S3	750
S5	250	750 μl S4	750
S6	125	750 μl S5	750
S7	62.50	750 μl S6	750
S8	31.25	750 μl S7	750
S9	15.63	750 μl S8	750
S10	7.81	750 μl S9	750
S11	3.91	750 μl S10	750
S12	1.95	750 μl S11	750

For serotonin, standards are prepared as follows:

	Serotonin (μM)	Volume (μl) Serotonin 20 mM	Volume (μl) Dilution Buffer
S1	1000.00	50	950
Serial dilutions			
S2	500.00	400	400
S3	250.00	400	400
S4	125.00	400	400
S5	62.50	400	400
S6	31.25	400	400
S7	15.63	400	400
S8	7.81	400	400
S9	3.91	400	400
S10	1.95	400	400
S11	0.98	400	400
S12	0.49	400	400

9.4.2.1 Adsorption of Dopamine–Albumin Conjugate

Prepare the dopamine–BSA in the coating buffer at a final concentration of 5 μg/L. Transfer 100 μL of this solution to each well (microplate for luminescence Microlite 2). Keep three wells without dopamine–BSA as blanks to control for nonspecific binding of antibodies. Place cover or seal with parafilm membrane and incubate overnight at 4°C in the dark. After the incubation period, remove all liquid by aspiration or by inverting the microplate on an absorbent towel. Wash wells with 200 µL of PBS three times. Do not dry wells. Immediately add 250 µL of blocking buffer in each well and incubate for 90 min at room temperature in the dark.

In the meantime, prepare the samples, dopamine standards (see above), and the primary antibody in dilution buffer. Remove the blocking buffer in wells and wash once with washing buffer. Mix 50 µL of each dopamine standard, sample, and blank (dilution buffer only) to 50 µL of primary antibody in wells containing adsorbed dopamine−BSA. Incubate at room temperature for 90 min in the dark. During that time, the primary antibody will bind proportionally to dopamine in the samples or S2−S11standards, i.e., the more dopamine in solution the less binding to the adsorbed dopamine−BSA at the bottom of the wells (competitive binding).

During this incubation step, prepare the dilution of the secondary antibody in dilution buffer. At the end of the incubation period, wash three times with 200 µL of washing buffer. Add 100 µL of secondary antibody conjugated to peroxidase enzyme. Cover and incubate for 60 min in the dark.

During that time, prepare the chemiluminescent substrate according to the supplier kit. An in-house preparation could be done by preparing 1−5 µM luminol with 50−100 µM H_2O_2 (prepare fresh just before the assay). After the incubation period, remove the secondary antibody and wash three times with washing buffer (PBS). Add 100 µL of chemiluminescent substrate and read luminescence after 3−5 min in a microplate reader (e.g., Chameleon Instrument, direct readings, 500 ms reading time).

9.4.3 Data Calculation and Analysis

A competitive binding curve is produced as shown in Figure 9.2. This curve will serve to extrapolate the levels of dopamine in the test sample. Because quantitation is based on inhibition of peroxidase activity, spiked samples should be tested to remove any possible matrix effects at first. If no interference is observed in the presence of the test sample, then dilutions of the test sample could be used to determine the concentration of dopamine within the linear equation in Figure 9.2. If interference is found, then calibration should be done by the standard addition methodology as explained in Chapter 8.

The extrapolated concentration of dopamine (nmol/mL) is then normalized against total proteins (mg/mL) or tissue homogenate weight (g/mL) to give nmol dopamine (or serotonin)/mg proteins.

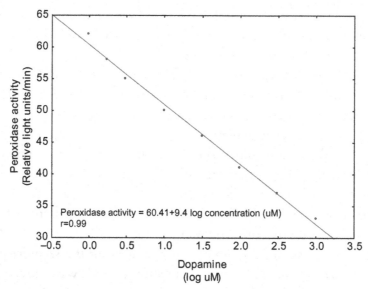

Figure 9.2 Typical competition curve of enzyme-based immune assays.

9.5 MONOAMINE OXIDASE ACTIVITY

Monoamine oxidase activity (MAO) is found in the mitochondria and the assay should be performed in the mitochondria fraction for maximum sensitivity. Mitochondria are prepared by first centrifuging the homogenate (prepared in isotonic conditions) at $1500 \times g$ for 15 min at $2-4°C$ and centrifuging the supernatant at $9000 \times g$ for 20 min at $2-4°C$. The pellet corresponds to the crude mitochondrial fraction and is resuspended in a small volume of isotonic buffer. However, the assay can be performed up to the homogenate fraction or supernatants obtained below $6000 \times g$ but cytosolic peroxidase would be present where specific controls should be included (by measuring in the presence/absence of the substrate tyramine). The principle of the assay is based on the detection of released H_2O_2 by the dichlorofluorescein diacetate/peroxidase system that is formed from the deamination reaction of biogenic amines (serotonin, dopamine) or nonspecific generic substrate such as tyramine.

9.5.1 Reagents

Assay buffer: Prepare 140 mM NaCl and 10 mM HEPES-NaOH, pH 7.4.
Tyramine substrate (1 mM): Dissolve 14 mg in 10 mL SQ water.

Dichlorofluorescein/peroxidase: Prepare a stock solution of dichlorofluorescein diacetate at 10 mM in dimethylsulfoxide (DMSO). Prepare a 5 mg/mL stock solution of horseradish peroxidase. Dilute the peroxidase stock solution to obtain 10 μg/mL in the assay buffer. Finally, dilute dichlorofluorescein diacetate at 10 μM with the 10 μg/mL peroxidase solution.

Aminotriazole (catalase inhibitor): Prepare a 10 mM stock solution, i.e., 84 mg in 10 mL SQ water.

Fluorescein standard: Prepare a 10 mM fluorescein in the assay buffer. Dilute 1/100 to obtain 100 μM fluorescein in the assay buffer. Dilute again at 1/100 to obtain 1 μM. The standard solution is prepared by mixing 20 μL of 1 μM fluorescein with 180 μL of the assay buffer.

9.5.2 Procedure

In a microplate for fluorescence, mix 40 μL of homogenate, mitochondria fraction (better) or homogenization buffer (blank), 10 μL of tyramine, 20 μL of dichlorofluorescein diacetate/peroxidase solution, 10 μL aminotriazole, and 20 μL of assay buffer. Aminotriazole prevents the hydrolysis of H_2O_2 by catalases. Incubate between 30 and 60 min at 30°C and take fluorescein readings at 485 nm excitation and 525 nm emission at each 10 min interval. In separate wells, a blank is prepared with only the supernatant, which might contain trace amounts of endogenous peroxidase and H_2O_2 without the addition of tyramine (replaced by the assay buffer). For calibration, a buffer blank and fluorescein standard 0.1 μM is prepared in the assay buffer.

9.5.3 Data Analysis

The enzyme activity is determined by the rate of change of fluorescein fluorescence units (FU) in time:

$$[FUsample - FUblank] \text{ at time } t \text{ min} - [FUsample - FUblank]$$
$$\text{at time } t\text{``0''}\text{min or blank} \times (1/\text{time } t \text{ min}) = \text{Change FU/min.}$$

FUs could be transformed into nmol fluorescein/mL by multiplying with the following:

[Fluorescein standard concentration/(FUstandard−FU blank)] to give nmol/(min/mL) fluorescein formed. The enzyme activity is then normalized against total proteins or tissue homogenate weight to obtain specific activity (nmol fluorescein/min/mg proteins or g tissue weight).

9.6 NEUROTRANSMITTER DIRECTED ADENYLATE CYCLASE ACTIVITY IN SYNAPTOSOMES

Synaptosomes consist of membrane vesicles of the synapse, which are rich in receptors (uptake) and transporters (reuptake) of neurotransmitters and sodium/potassium pumps to maintain transmembrane polarity [14]. For example, serotonin reuptake inhibitors used to treat depression inhibit the reuptake of serotonin in synapse. During that process, the serotonin transporter exchanges one potassium ion for one serotonin molecule, which activates the membrane Na/K-ATPase activity to maintain the electrochemical gradient [15]. Activation of receptors under the control of GTP/adenylate cyclase could also be measured using this system. For example, opiate peptides in the nerve ganglia inhibited dopamine-stimulated Na/K-ATPase in the blue mussel [16]. Hence, synaptosomes represent an interesting neurophysiological model to study neuroendocrine effects of environmental contaminants.

9.6.1 Reagents

Homogenization buffer: Prepare 250 mM sucrose (0.5 L) containing 10 mM HEPES-NaOH, pH 7.4, 1 mM EDTA, and 1 mM dithiothreitol (as a reducing agent). The buffer solution is stable for 2 weeks at 4°C.

0.8 M sucrose: Prepare a 0.8 M sucrose solution containing 1 mM HEPES-NaOH, pH 7.4, and 0.1 mM EDTA.

Assay buffer: Prepare 10 mM HEPES buffer, pH 7.4, containing 140 mM NaCl, 1 mM $MgCl_2$, and 1 mM KCl.

Ascorbate: Prepare 1% solution (0.1 g/10 mL water) daily.

Molybdate: Prepare 1% ammonium molybdate in 0.9 M H_2SO_4.

ATP stock: Prepare daily 5 mM ATP in SQ water.

Dopamine and serotonin: Prepare daily in separate tubes 5 mM serotonin and dopamine in SQ water.

Reaction media: Mix 1 mL ATP stock, 1 mL of either serotonin or dopamine, and 48 mL of assay buffer before the start of the assay.

Trichloroacetic acid: Prepare a 1% solution in SQ water in glass bottle.

Phosphate standard: Prepare a 0.68 mM sodium phosphate standard in SQ water.

9.6.2 Procedure

Synaptosomes are prepared by a convenient sucrose density methodology although other available methods are commercially available. Brain and nerve tissues are homogenized in the homogenization buffer using a Teflon pestle tissue grinder apparatus. Homogenization is complete with four to six passes on ice. The material is then centrifuged $1500 \times g$ for 20 min at 2−4°C. The supernatant is collected and centrifuged at $9000 \times g$ for 20 min at 2−4°C. The pellet is removed from the supernatant and resuspended in the homogenization buffer and centrifugation at $9000 \times g$ for 20 min. The pellet is resuspended in a small volume (0.25−0.5 mL) of the homogenization buffer and layered on top of 0.5−1 mL of 0.8 M sucrose. Centrifuge at $9000 \times g$ for 20 min at 2−4°C, quickly recuperate the supernatant (the pellet contains crude mitochondria), add 2 volumes of SQ water to restore osmolarity, and keep on ice or store at −85°C until analysis.

Na/K-ATPase activity is determined by determining the hydrolysis of ATP in the presence of synaptosomes and either added dopamine or serotonin. Addition of other xenobiotics could also be added to study interactions in synaptosomes activity. The released phosphates from ATP hydrolysis are determined by the colorimetric phosphomolybdate assay (see ALP assay). Pre-mix 200 μL of synaptosome in 800 μL of reaction media (with or without serotonin or dopamine at concentrations between 10−100 μM). Transfer 100 μL of the mixture in a clear microplate and incubate for 30 min at 20−25°C. The reaction is stopped be the addition of 5 μL of TCA. The blank consists of the $t = 0$ min, i.e., 100 μL of the reaction media with synaptosomes that was immediately added to TCA in a separate microplate well. A standard is also prepared in another well containing the blank ($t = 0$ min) with 10 μL of phosphate standard (4 μM final concentration).

Total free phosphate is determined using the phosphomolybdate assay. In a clear microplate (containing the samples), add 25 μL of molybdate, 25 μL ascorbate, and 45 μL of water. Mix for 5 min and read absorbance at 815 and 444 nm (optional for the latter).

9.6.3 Data Calculation and Analysis

The phosphomolybdate complex absorbs at 815 nm. The absorbance at 444 nm could be measured if turbidity is present in the wells to correct for background absorbance as follows [6]:

$$\text{Corrected A815 nm} = (1.045 \times \text{A815}) - (0.043 \times \text{A444}).$$

Na/K-ATPase activity is determined as follows:

$$(\text{A815 corr 30 min} - \text{A815 blank corr}) \times 5 \text{ μM phosphate}/(\text{A815 standard}$$
$$- \text{A815 blank}) \times 1/30 \text{ min} \times \text{dilution factor } (200/10 \times (1000/200))$$
$$= \text{liberated phosphates/min/ml}.$$

The enzyme activity is normalized against total proteins of the isolated synaptosomes to obtain Na/K-ATPase activity: ATP hydrolysis/min/mg proteins.

9.7 ACETYLCHOLINESTERASE

Acetylcholine is a neurotransmitter involved in muscle tone and contraction and its levels are regulated by acetylcholinesterase (AChE), which hydrolyzes the acetate group to form choline: acetylcholine → choline + acetate [16]. AChE was and still is a commonly used biomarker in aquatic ecotoxicology studies. At first, organophosphate, organochloride, and carbamate pesticides were designed as inhibitors of this enzyme leading to sustained muscle contraction, paralysis, and death because of the decoupling between loss of enzyme activity and the increased half-life of acetylcholine in tissues. Indeed, this enzyme was extensively used as a biomarker of exposure of these pesticides in nontarget organisms such as fish [17], mussels [18], and sea urchins [19]. More recently, this enzyme was used as a marker of neural activity in *Artemia* shrimp, freshwater mussels [20,21], and amphipods [22]. Changes in this enzyme activity could lead to altered behavior and neuromuscular activity in organisms and could serve as a biomarker of neurotoxicity in organisms. However, this enzyme could be misused in some cases as a surrogate of acetylcholine levels and precautions/validation should be taken to ensure whether changes in enzyme activity are correlated with acetylcholine levels. For example, a decrease in enzyme activity could be the result of reduced acetylcholine signaling or inhibition of the enzyme by xenobiotics (organophosphorus pesticide), which has different physiological consequences.

9.7.1 Reagents

Tris-acetate buffer: Prepare a 100 mM solution at pH 7.2. Dissolve 1.21 g of TRIS-base in 90 mL of water, adjust pH to 7.4 with 10% acetic acid, and complete to 100 mL with water. A 50 mM Tris-acetate buffer could be prepared by diluting with one volume of SQ water.

Acetylthiocholine reagent: Prepare a 1 mM working solution in 50 mM Tris-acetate buffer. (9.9 mg in 50 mL of 50 mM Tris-acetate buffer).

Ellman's reagent: Prepare a 5 mM stock solution of DTNB in phosphate-EDTA buffer. Dissolve 1.36 g of KH_2PO_4 and 4.5 mg EDTA in 80 mL of bidistilled water. Adjust pH to 7.2 with NaOH and add 0.2 g DTNB.

Complete to 100 mL with water. This stock solution is stable for 2 months in the dark. Ellman's reagent is prepared by diluting five times the DTNB stock solution in 100 mM Tris-acetate buffer.

Thiol standard: Prepare a 1 mM stock solution by dissolving 3 mg of reduced GSH in 10 mL of 0.1 N HCl. Stable for 1 week in the dark at 4°C. Prepare a 25 μM standard solution by diluting 0.25 mL of the GSH stock in 10 mL of Tris-acetate buffer. Prepare daily.

9.7.2 Procedure

Tissues are homogenized using a Teflon tissue pestle in 10 mM Tris-acetate, pH 7.4, containing a protease inhibitor (10 μg/mL apoprotinin), 140 mM NaCl, and 1 mM EDTA. Any other homogenization buffer could be used. The tissue is homogenized using a Teflon pestle tissue grinder (four to six passes) on ice and 1 mL of the homogenate is weighted. The homogenate is then centrifuged at 12,000 × *g* for 20 min at 2−4°C. The supernatant (S12) is removed with a pipette and the upper lipid layer discarded. The S12 fraction could be stored at −85°C.

In a clear 96-well microplate, mix 50 μL of tissue extract (S12) with 100 μL of 1 mM acetylthiocholine and 100 μL of Ellman's reagent. Incubate at 30°C for 0, 5, 10, 20, and 40 min and measure the absorbance at 412 nm at each time. The blank is *t* − 0 min or the S12 fraction from which no acetylthiocholine was added (replaced by 100 μL buffer). Calibration is achieved by preparing a 25 μM thiol standard in 250 μL total volume. The analytical blank is the Tris-acetate buffer.

9.7.3 Data Analysis

The increase in absorbance from the reaction between the Ellman's reagent and thiocholine is measured within the linear portion in time as follows:

$$[(A412 \text{ sample} - A412 \text{ blank}) 20 \text{ min} - (A412 \text{ sample} - A412 \text{ blank}) 0 \text{ min}]$$
$$\times 1/20 \text{ min} \times (25 \text{ nmol/mL})/(A412 \text{ standard} - A412 \text{ blank})]$$
$$= \text{nmol of thiocholine formed}/(\text{min} \times \text{mL}) \times \text{dilution factor } (250/50)$$

This value is then normalized against total proteins in the S12 (mg/mL) or homogenate weight to give nmol thiocholine formed/(min × mg proteins or g).

9.8 STEROID ANALYSIS BY HIGH-PERFORMANCE THIN LAYER CHROMATOGRAPHY

The capacity of gonad homogenates to produce sexual steroids (testosterone and estradiol-17β) could be determined by using high-performance thin layer chromatography (HP-TLC). This approach could replace steroid immunoassays assessment as shown previously, although it is less sensitive and could be used to determine total cholesterol levels in tissue extracts. HP-TLC has increased sensitivity and is more amenable to quantification because the peaks are compact (forms a thin line) and not smeared (spots) as with normal thin layer chromatography. This assay could also be used to measure steroid production in gonad tissues responsible for steroidogenesis. Gonad homogenate extracts are exposed *in vitro* to testosterone or progesterone and NADPH where the formation of estradiol-17β or testosterone could be, respectively, followed by HP-TLC image analysis. This technique could also be used to determine the presence of other compounds such as coprostanol (reduced products from bile acids by gut microflora, a marker of fecal matter contamination), nonylphenol, and general steroid analysis in various organisms. The assay for estradiol-17β biogenesis in testosterone-spiked gonad is presented as a marker of aromatase activity.

9.8.1 Reagents

Homogenization buffer: Prepare 125 mM NaCl, 10 mM KH_2PO_4, pH 7.4, 1 mM EDTA, and 1 mM dithiothreitol. Stable for 4 weeks at 4°C.

Assay buffer: Prepare 100 μM testosterone (or progesterone for testosterone production) and 0.1 mM reduced NADPH in 10 mM KH_2PO_4, pH 7.4, containing 125 mM NaCl and 1 mM $NaHCO_3$.

HCl: Prepare 1 M from concentrate (8 M) in water: 1.2 mL of concentrated HCL in 10 mL water (fume hood, wear protective clothing, gloves, and eyewear).

HP-TLC plate: Silica-based HP-TLC for maximum resolution and sensitivity (Sigma Chemical Company or Merck) measuring 8 × 10 cm.

Resolving solvent: Prepare 60% hexane and 40% ethyl acetate in a fume hood.

Phosphomolybdate spray: 1−10% phosphomolybdate in ethanol 100%.

9.8.2 Procedure

The gonad tissues are collected and weighted and a portion kept aside for histological analysis (sex identification and state of gonad maturation). The gonad is then ground using a Teflon pestle tissue grinder in 5 volumes of ice-cold homogenization buffer. The homogenate is centrifuged at $12,000 \times g$ for 30 min at 2°C. The supernatant (S12) is kept on ice.

The S12 fraction is added to 5 volumes of assay buffer and allowed to incubate for 0, 60, and 120 min at 25°C. The reaction is stopped by the addition of 0.2 volume of HCl 1 M and placed on ice for 30 min to permit the hydrolysis of any steroid conjugate(s). Two volumes of ethyl acetate are added to the reaction mixture and mixed and centrifuged briefly at $3000 \times g$ for 2 min to separate the phase. The ethyl acetate phase is then removed and evaporated under nitrogen stream. The material is suspended in 50–100 μL ethyl acetate and spotted 5×10 μL (air dried between applications) on the HP-TLC plate. The plates are then placed in chambers containing 15 mL of resolving solvent. After migration of the solvent for at least 80% of the plate height, the plates are removed and dried in a fume hood. The plates are then sprayed with phosphomolybdate spray and dried at 80–100°C in an oven. If no oven is available then a hand-held air dryer could be used. The spots (Figure 9.3) are scanned with a conventional scanner and the intensity of the bands analyzed by image analysis software (Un-Scan-It gel analysis). Another means of quantitation consists of scraping the silica powder off the stained band and eluting with ethanol and measuring the optical density at 660–810 nm (absorption peak of reduced phosphomolybdate complex).

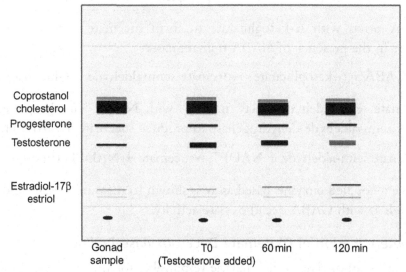

Figure 9.3 Representation of an HP-TLC chromatogram for sexual steroid production.

For calibration, standard solutions of testosterone, cholesterol, progesterone, and 17β-estradiol at 100 μM concentration in ethanol are included in a separate lane either together or in different lanes at first to validate good separation between them. The band densities are then taken for calibration.

9.8.3 Data Analysis

The band intensity is determined and compared with a standard solution of estradiol-17β and testosterone and resolved in the HP-TLC plates. The amounts in nanograms of estradiol-17β are then converted to concentrations by dividing with the applied volume (50 μL) to give the concentration ng estradiol-17β/mL. The concentration is then normalized against the total protein content of the S12 fraction or the tissue density (g/mL) to give ng/mg proteins or ng /g tissues.

9.9 EVALUATION OF GLUTAMATE AND γ-AMINOBUTYRATE

γ-Aminobutyrate (GABA) is a neuroinhibitory transmitter that involves two types of receptors, GABA1 and GABA2, which are ionotropic and metabotropic receptors for action. The latter form calls for signal transduction (coupled to G-protein). It is produced from glutamate, which is considered, conversely, a neuroexcitatory neurotransmitter. GABA is produced from the decarboxylation of glutamate catalyzed by glutamate decarboxylase:

$$\text{L-glutamate} + \text{pyridoxal 5-phosphate} \rightarrow \text{GABA} + \text{CO}_2.$$

GABA reacts with α-ketoglutarate to form succinate semi-aldehyde and glutamate in the presence of GABA transaminase:

$$\text{GABA} + \alpha\text{-ketoglutarate} \rightarrow \text{succinate semi-aldehyde} + \text{Glutamate}$$

Succinate semi-aldehyde reacts, in turn, with $NADP^+$ in the presence of succinate semi-aldehyde dehydrogenase to produce succinate + NADPH.

$$\text{Succinate semi-aldehyde} + NADP^+ \rightarrow \text{succinate} + \text{NADPH (fluorescent)}$$

Hence a coupled enzyme-based assay is shown to measure GABA and glutamate levels as with GABA decarboxylase activity.

9.9.1 Determination of Glutamate Decarboxylase Activity

Glutamate decarboxylase is the enzyme responsible for the conversion of glutamate into GABA, which is an inhibitory neurotransmitter. The assay is based on the multi-enzymatic reaction that involves GABA transaminase and succinate semi-aldehyde dehydrogenase.

9.9.1.1 Reagents

Sodium phosphate buffer (1 L): Prepare 0.1 M NaH_2PO_4 and adjust pH to 7 with NaOH.

Glutamate stock solution: Prepare 0.5 M glutamate in 1 M NaOH: 0.7 g glutamic acid in 10 mL of NaOH 1 M (might require some heating at 40−50°C for 5 min).

Stock AET 10×: Prepare 10 mM solution of 2-aminoethylisothironium bromide (AET) in 0.1 M KH_2PO_4, pH 7 (sodium phosphate buffer).

Stock pyridoxal 5-phosphate (PP; 50×): Prepare 1 mM solution (100 mL) in sodium phosphate buffer: 25 mg in 100 mL. Stable for 1 month when kept at 4°C in the dark.

Assay buffer: Prepare 5 mL of AET (10×), 1 mL PP (50×), 0.05 g Triton X-100 and complete to 50 mL with 0.1 M sodium phosphate buffer.

GAD substrate: Add 10 mL of glutamate 500 mM to 25 mL of PP. Add 0.1 g of dithiothreitol and complete to 100 mL with 0.1 M sodium phosphate buffer.

9.9.1.2 Procedure

Prepare seven tubes denoted Rx, T0, T15, T30, T45, T60, and endogenous GABA. Preheat 2 water baths: one at 30°C and the other at 90°C.

Dilute the S12 fraction ½ in assay buffer in the Rx tube—60 μL of S12 and 60 μL of assay buffer.

Start the reaction by adding 120 μL of GAD substrate. Incubate at 30°C for 0, 15, 30, 45, and 60 min. At each time, transfer 40 μL of the reaction mix in the corresponding tube and add 20 μL of 0.2 N HCl. Heat for 5 at 100°C to destroy endogenous NADPH. Keep at −80°C for GABA assay (see below).

9.9.2 Determination of Endogenous GABA

In microcentrifuge tubes, mix 20 μL of S12 with 20 μL of sodium phosphate buffer and 20 μL of 0.3 N HCl. Heat 5 min at 90−100°C to remove endogenous NADPH. Keep at −80°C for GABA assay.

Note: The levels of endogenous glutamate could be determined by running the assay without the addition of glutamate for 60 min to convert the endogenous glutamate into GABA. However, glutamate could be formed from endogenous α-ketoglutarate by α-ketoglutarate dehydrogenase in the presence of NADH. If GABA decarboxylase is (commercially available) added then the S12 fraction could be spiked by the enzyme in excess to force the complete conversion of glutamate into GABA.

9.9.3 Detection of GABA

The assay is based on the reaction of GABA with α-ketoglutarate leading to the formation of succinate semialdehyde and glutamate (GABA transaminase or GABase). Succinate semialdehyde then is oxidized with NADP to form succinate and NADPH, which is detected by fluorescence.

9.9.3.1 Reagents

A. *Assay buffer*: Prepare 300 mM Tris-HCl at pH 8.4 (1 L).

B. *α-Ketoglutarate solution*: Prepare 100 mM in SQ water (10 mL).

C. *GABA transaminase/succinate semi-aldehyde dehydrogenase*: Dilute the commercial GABase stock powder (includes succinate semi-aldehyde dehydrogenase) at 3.35 U/mL (5 mg/mL) in assay buffer adjusted with the pH readjusted to 7.4 with HCl. Prepare 100 μL aliquots and store for 1 month at −20°C.

D. *Stock dithiothreitol*: Prepare 100 mM solution in SQ water (5 mL).

E. *Stock NADP$^+$*: Prepare 25 mM solution in SQ water (10 mL). Prepare 10 μL aliquots and store for 1 month at −20°C.

F. *Stock GABA*: Prepare a 10 mL 25 mM solution in the assay buffer (A).

9.9.3.2 Procedure

Prepare daily the detection mixture (10 mL) by mixing 8.7 mL of solution A, 1 mL of solution B, 0.1 mL of solution D, and 0.2 mL of solution E. Mix well and remove 1 mL for the blanks (no enzyme). Preheat the microplate reader sample rank at 30°C if available.

In a 15 mL tube, mix 8.83 mL of the above solution and add 270 μL of solution C. Final conditions: 300 mM Tris-HCl, pH 8.4, 10 mM α-ketoglutarate, 1 mM dithiothreitol, 0.5 mM NADP$^+$, and 100 mU/mL GABase.

In a clear microplate, add 10 μL of sample (GABA formed and endogenous GABA), blank, and standard (0.1 mM GABA to give 100 nmol/mL final concentration). Start the reaction by adding 150 μL of the detection mixture and read fluorescence (formation of NADPH) at 350 nm excitation and 450 nm emission at 0 and 20 min. If matrix effects are suspected in the sample, then the standard is spiked to the sample to provide an internal standard.

9.9.3.3 Data Analysis

The enzyme activity is determined by the following relationship:

NADPH fluorescence (Sample − blank) × [20 nmol/mL GABA/

NADPH fluorescence (standard − blank)] × 1/incubation time

× dilution factor (160/10 × 60/40 × 240/60) = nmol GABA formed/min/mL.

The activity is then normalized against total proteins in the S12 fraction or homogenate weight to give nmol GABA/min/mg proteins or g tissues.

REFERENCE

[1] Gagné F, Blaise C. Effects of municipal effluents on serotonin and dopamine levels in the freshwater mussel *Elliptio complanata*. Comp Biochem Physiol 2003;136C:117−25.

[2] Waye A, Trudeau VL. Neuroendocrine disruption: more than hormones are upset. J Toxicol Environ Health 2011;14B:270−91.

[3] Gagné F, André C. New approaches to indirect vitellogenin-like protein evaluations in aquatic oviparous and ovoviviparous organisms. Fresenius Environ Bull 2011;20:12−7.

[4] Hallgren P, Matensson L, Mathiasson L. Improved spectrophotometric vitellogenin determination via alkali-labile phosphate in fish plasma—a cost effective approach for assessment of endocrine disruption. Intern Environ Anal Chem 2009;89:1023−42.

[5] Hallgren P, Martensson L, Mathiasson L. A new spectrophotometric method for improved indirect measurement of low levels of vitellogenin using malachite green. Int J Environ Anal Chem 2012;92:894−908.

[6] Stanton MG. Colorimetric determination of inorganic phosphate in the presence of biological material and adenosine triphosphate. Anal Biochem 1968;22:27−34.

[7] Byrne BM, Gruber M, Ab G. The evolution of egg yolk proteins. Prog Biophys Mol Biol 1989;53:33−69.

[8] Gagnaire B, Gagné F, André C, Blaise C, Abbachi K, Budzinski H, et al. Development of biomarkers of stress related to endocrine disruption in gastropods: alkali-labile phosphates, protein-bound lipids and vitellogeninlike proteins. Aquat Toxicol 2009;92:155−67.

[9] Gagné F, Bouchard B, André C, Farcy E, Fournier M. Evidence of feminization in wild Elliptio complanata mussels in the receiving waters downstream of a municipal effluent outfall. Comp Biochem Physiol 2010;153C:99−106.

[10] Gagné F, Blaise C, Salazar M, Hansen P. Evaluation of estrogenic effects of municipal effluents to the freshwater mussel *Elliptio complanata*. Comp Biochem Physiol C Toxicol Pharmacol 2001;128:213−25.

[11] Peuler JD, Passon PG. An automated assay for brain serotonin. Anal Biochem 1973;52:574−83.

[12] Orsinger OA, Marichich ES, Molina VA, Ramirez OA. A reliable and sensitive method for the simultaneous determination of dopamine, noradrenaline, 5-hydroxytryptamine and 5-hydroxy-indol acetic acid in small brain samples. Acta Physiol Latino-America 1980;30:111−5.

[13] Kim J, Jeon M, Paeng KJ, Paeng IR. Competitive enzyme-linked immunosorbent assay for the determination of catecholamine, dopamine in serum. Anal Chim Acta 2008;619:87−93.

[14] Dunkley PR, Jarviel PE, Robinson PJ. A rapid Percoll gradient procedure for preparation of synaptosomes. Nat Protoc 2008;3:1718—28.

[15] Lajeunesse A, Gagnon C, Gagné F, Louis S, Cejka P, Sauvé S. Distribution of antidepressants and their metabolites in brook trout exposed to municipal wastewaters before and after ozone treatment. Evidence of biological effects. Chemosphere 2011;83:564—71.

[16] Ellman GL, Courtney KD, Andres Jr V, Featherstone RM. A new and rapid colorimetric determination of acetylcholinesterase activity. Biochem Pharmacol 1961;961:88—95.

[17] Szeto SY, Sundaram KM, Feng J. Inhibition of brain AChE in brook trout by aminocarb and its toxic metabolites. J Environ Sci Health B 1985;20:559—75.

[18] Stefano GB, Leung M. Purification of opioid peptides from molluscan ganglia. Cell Mol Neurobiol 1982;2:347—52.

[19] Cunba I, Garcia LM, Guilhermino L. Sea-urchin (*Paracentrotus lividus*) glutathione S-transferases and cholinesterase activities as biomarkers of environmental contamination. J Environ Monit 2005;7:288—94.

[20] Gagné F, Blaise C. Shell protein characteristics and vitellogenin-like proteins in brine shrimp *Artemia franciscana* exposed to municipal effluent and 20-hydroxyecdysone. Comp Biochem Physiol 2004;138C:515—22.

[21] Gagné F, André C, Gélinas M. Neurochemical effects of benzodiazepine and morphine on freshwater mussels. Comp Biochem Physiol 2010;152C:207—14.

[22] Xuereb B, Lefèvre E, Garric J, Geffard O. Acetylcholinesterase activity in *Gammarus fossarum* (Crustacea Amphipoda): linking AChE inhibition and behavioural alteration. Aquat Toxicol 2009;94:114—22.

CHAPTER 10

Genotoxicity

François Gagné with the contribution of Émilie Lacaze, Sylvie Bony and Alain Devaux

Chapter Outline

Biochemical Ecotoxicology
DOI: http://dx.doi.org/10.1016/B978-0-12-411604-7.00010-6

10.1 INTRODUCTION

Exposure to compounds stemming from anthropogenic activity could lead to alterations in the genetic material. DNA damage arises from the primary modification of DNA, which could lead to altered protein function, inactivation, and genetic mutations. If gene mutations occur at a critical gene responsible for cell differentiation, communication, and cell growth, then there is the likelihood of producing transformed cells leading to altered cell growth, function, and up to cancers. Epigenetic mechanisms are also at play, which could lead to important changes in the relative balance between gene expression and regulation. It is thought that the production of DNA damage and continued DNA repair activities form the basis of carcinogenesis and gene mutations that could compromise the normal function of cells. Ideally, the analysis of specific DNA adducts should be realized, but this approach is not yet available to the plethora of chemicals in the environment with the exception of DNA-polyaromatic hydrocarbon (PAH) adducts, which can be determined using ^{32}P-post labeling or PAH-specific fluorescence of isolated DNA macromolecules. In the future, it is expected that methods based on HPLC-mass spectrometry will permit the analysis of nucleotide adducts. At present, the area of DNA damage assessment is much more amenable to the common laboratory in environmental research. These methods determine DNA damage at the molecular and cellular levels and provide insight into the level of DNA-induced damage and repair by environmental compounds.

10.2 ALKALINE PRECIPITATION ASSAY

Damaged DNA undergo repair by various repair enzymes such as endonucleases and exonucleases (DNA ligase, oxoguanine glycosylase, uracil nucleotide glycosylase, and APEX, to name a few), which produce single- and double-stranded DNA. Since these strands are usually protein free compared to genomic DNA, which is associated with proteins in the chromatin structure, the DNA strands could be easily separated by centrifugation in the presence of potassium dodecyl sulfate. A simple way to detect the presence of DNA strand breaks in cells was developed based on the separation of single- and double-strand breaks from genomic DNA [1]. Initially the method called for radioactively labeled nucleic acids for detection, but further modification permitted fluorometric detection in the presence of trace amounts of detergents [2,3]. This method represents a simple and rapid way to evaluate the genotoxic potential of miscellaneous compounds in cell or tissue homogenates. It could be used on either cryopreserved or freshly prepared tissues, hence there is no need to work with fresh tissues.

10.2.1 Reagents

DNA assay buffer: Prepare a 0.4 M NaCl solution containing 4 mM sodium cholate and 100 mM Tris base in 900 mL of SQ water, adjust pH to 8.5 with acetic acid, and complete at 1 L with water. Store in the dark at 4°C (stable for 4 weeks).

TE buffer: Prepare 10 mM Tris base buffer containing 10 mM EDTA in 90 mL of SQ water, adjust pH to 8 with 10% acetic acid, and complete to 100 mL.

DNA stock: Dissolve 10 mg of salmon sperm DNA (Sigma Chemical Company) in 90 mL of TE buffer. Heat at 70°C for 10−60 min for dissolution. Dilute 1/20 in TE buffer to obtain 50 μg/mL and store in 50−100 μL aliquots at −20°C.

DNA detection reagent: Prepare 1 mg/mL of Hoechst reagent in methanol. Dilute 1/100 in DNA assay buffer at the day of analysis. The alternate dye Pico Green (at 1/200 dilution) could be used also with excellent sensitivity if an UV excitation spectrometer is not available (requires 485 nm excitation/530 nm emission).

Sodium dodecyl sulfate (SDS) reagent: Prepare a 2% SDS containing 10 mM EDTA, 10 mM Tris base, and 50 mM NaOH in 100 mL of SQ water.

KCl reagent: Dissolve 1.2 g KCl in 100 mL of SD water (1.2%).

10.2.2 Procedure

The assay could be done directly on cells on tissue homogenates before or after conservation at −80°C. The tissue homogenate should be prepared with care as to minimize DNA shearing (no ultrasound, no Polytron). A Teflon pestle tissue grinder is recommend with three to five passes at 4°C. With cultured or sampled cells, no homogenization steps are required.

Add 250 µL of SDS reagent to 25 µL of cell suspension or tissue homogenates, mix by inversion, and add 1 volume of KCl reagent. Mix by inversion (no pipetting) and incubate at 60°C for 10 min. Place on ice for 5 min and centrifuge at $8000 \times g$ for 10 min. Carefully collect the supernatant to avoid resuspending the SDS pellet (containing genomic DNA) and place on ice.

To 25–50 µL of blank, standard, and sample, add 150 µL of DNA assay buffer and 50 µL of detection reagent. For the standard, mix 5 µL of DNA stock solution with 145 µL DNA assay buffer with 50 µL of detection reagent.

10.2.3 Data Handling and Calculation

The relative fluorescent values are converted to µg DNA equivalents using the salmon sperm DNA standard:

$$[\text{Fluorescence sample} - \text{blank}] \times \text{Standard concentration}/$$
$$[\text{Fluorescence standard} - \text{blank}] \times (\text{total volume for DNA assay}/$$
$$\text{sample volume µL}) \times (\text{total volume of SDS} - \text{KCl}/$$
$$\text{volume of added homogenate}) = Y \text{ µg/mL}$$

This concentration is then normalized against total proteins (mg/mL) or homogenate tissue weight in g/mL:

$$Y \text{ µg/mL/g tissue weight/mL or mg/mL protein} = Z \text{ µg/g or } Z \text{ µg/mg proteins}$$

The concentration of the standard solution should be close to the signal value of the sample, i.e., within $\pm 20\%$ of the fluorescence signal. A standard curve could also be constructed to convert the fluorescence values into DNA equivalents using blank samples to prepare the standards.

10.3 COMET ASSAY

Contributed by Émilie Lacaze et al.

10.3.1 Introduction

The Comet assay is a relatively simple, sensitive, and quantitative method to study DNA damage (including oxidative damage) and repair at the level of the individual cell. Initially developed by Singh et al. [4], this assay is now applicable to almost any eukaryotic cells, customizable depending on the goal of the user. Because of its flexibility, sensitivity, and ease of use, the Comet assay has become one of the most commonly used methods for studying DNA damage in single cells [5]. It is used in a wide scope: from regulatory safety assessment chemicals regarding genotoxicity to DNA damage and repair mechanistic studies and from human biomonitoring to ecogenotoxicology.

The Comet assay, also called single-cell gel electrophoresis, is based on agarose-embedded DNA electrophoresis. This method is adapted to determine DNA damage in isolated cells, so the cells need to be isolated from tissues first. DNA is a highly supercoiled macromolecule and negatively charged at normal pH; hence, under an electrophoretic field, DNA molecules migrate toward the anode. In isolated cells following a lysis step for removing cell and nuclear membranes and all cytoplasmic components, supercoiled nuclear DNA forms a nucleoid, a kind of DNA ball similar to a nucleus in shape. Deprived of membranes and histones, DNA breaks in the nucleoid, leading to a more relaxed DNA core. The stretching of DNA-relaxed loops toward the anode during electrophoresis will form a comet-like structure with a head and a tail (Figure 10.1). DNA loops/fragments constitute the comet tail since they migrate more quickly than the nucleoid DNA core during electrophoresis. After DNA staining with a fluorescent dye, comets are visualized with an epifluorescence microscope. The fluorescence intensity in the comet tail indicates the extent of DNA damage.

DNA can be subjected to several kinds of damage. Examples of the various types of DNA damage, their main causes and repair mechanisms involved are summarized in Figure 10.2. The Comet assay allows for the detection of a wide array of DNA damage in single cells with a high sensitivity. It detects DNA strand breaks (single and double stranded) as well as incomplete excision repair and cross-linking sites. During the assay, unwinding of DNA at alkaline pH (>13) enables the hydrolysis of alkali-labile sites to DNA breaks. Alkali-labile sites include apurinic/apyrimidinic sites, baseless sugars resulting from spontaneous loss of bases or as intermediates in base excision repair (in which bases are removed by glycosylases). Intercalating agents generate DNA unwinding, therefore, single-stranded loops could also be detected by the Comet assay.

The sensitivity of the assay depends on its calibration and there are some variations between laboratories, although substantial efforts have been made to standardize the assay. It is generally accepted that the frequency of DNA damage detectable with the Comet assay is from 0.06 to 3 breaks per 10^9 Da of genomic DNA. Roughly speaking one hundred to several thousand breaks per cell can be determined [6].

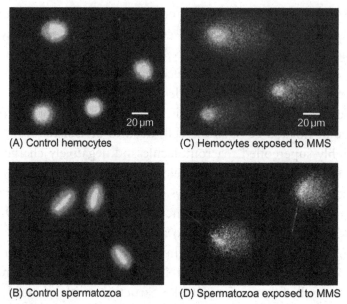

(A) Control hemocytes   (C) Hemocytes exposed to MMS

(B) Control spermatozoa   (D) Spermatozoa exposed to MMS

Figure 10.1 Comet images from *Gammarus* cells (hemolymph and sperm tissues) nonexposed (A and B) and exposed *in vitro* for 1 h to the model genotoxicant methyl methanesulfonate (C and D). The figure shows that nucleoids remain similar in shape as the nucleus [5].

10.3.2 Reagents and Materials

Chemical Products	Abbreviation
Dimethylsulfoxide	DMSO
Disodium EDTA	Na_2EDTA
Absolute ethanol	EtOH
Low melting point agarose	LMPA
Normal melting agarose	NMA
Phosphate buffered saline $Ca^{2+}Mg^{2+}$ free, pH 7.5	PBS
Sodium chloride	NaCl
Sodium hydroxide	NaOH
SYBR Green I (or ethidium bromide)	
Triton X-100	
Trizma base	Tris

Specific materials

Coplin jars	
Slide staining tray	
Horizontal gel electrophoresis apparatus	
Epifluorescence microscope	

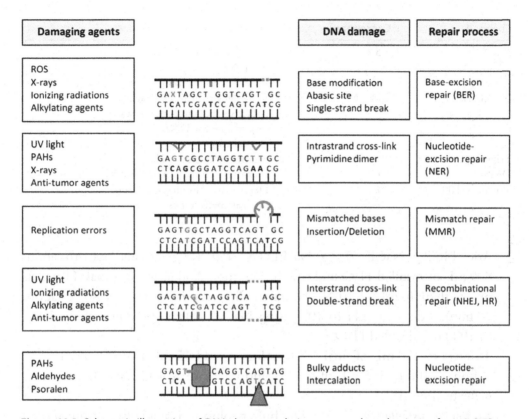

Figure 10.2 Schematic illustration of DNA damages, their causes, and mechanisms of repair [35].

NMA: Prepare NMA 1% in distilled water: 1 g NMA per 100 mL distilled H₂O. Heat near boiling (microwave can be used) in a glass beaker until the agarose completely dissolves.

LMPA: Prepare LMPA 1% (500 mg LMPA per 50 mL PBS). Aliquot 5 mL samples into vials. Store at 4°C until needed. When needed, briefly melt agarose in microwave or by another appropriate method. Place LMPA vial in a 37°C dry or water bath to cool and stabilize the temperature.

Lysing solution: A stock solution without DMSO and without Triton X-100 can be prepared and stocked at 4°C for months. The final lysing solution with DMSO and Triton X-100 must be prepared extemporaneously.

Chemical	Concentration	Purpose	
NaCl	2.5 M	Removes histones from the DNA	⎫
		Produces an osmotic shock	⎪
Na$_2$EDTA	100 mM	Scavenges radicals and metallic cofactors to inhibit nucleases, responsible for DNA degradation	⎬ Stock solution
Trizma base	10 mM	Maintains alkaline pH	⎭
DMSO	10%	Increases cell membrane fluidity	
Triton X-100	1%	Detergent, solubilizes cell and nuclear membranes	

Add 146.2 g NaCl, 37.2 g Na$_2$EDTA, and 1.2 g Trizma base to about 800 mL of distilled H$_2$O and stir the mixture. Add about 8 g NaOH pellets or less than 40 mL of 10 N NaOH to allow the mixture to dissolve (about 30 min). Adjust the pH to 10.0 using NaOH pellets and adjust the volume to 1000 mL distilled H$_2$O.

To prepare 150 mL of final lysing solution, add 15 mL DMSO and 1.5 mL Triton X-100 to 135 mL of stock solution. Store at 4°C for at least 30 min prior to use.

Electrophoresis buffer, pH > 13: The buffer solution is made fresh from several stock solutions before each electrophoresis run. The total volume of electrophoresis buffer to be prepared depends on the electrophoresis tank capacity and must be evaluated prior to the experiment. Prepare 5 N NaOH stock solution: 200 g NaOH/1000 mL distilled H$_2$O. Store at room temperature. Prepare also 200 mM Na$_2$EDTA stock solution by dissolving 14.9 g Na$_2$EDTA in 200 mL distilled H$_2$O. Store at 4°C. To prepare 1000 mL of electrophoresis buffer, add 60 mL NaOH stock solution and 5 mL Na$_2$EDTA stock solution to distilled H$_2$O. Store at 4°C for at least 30 min prior to use.

Neutralization buffer: Prepare 0.4 M Tris neutralization buffer as follows: add 48.5 mg of Tris to 800 mL distilled H$_2$O, adjust pH to 7.5 with concentrated HCl, and complete to 1000 mL with H$_2$O. Store at 4°C.

Staining solution: Dilute SYBR Green stock solution (SYBR Green 10,000 × in DMSO) 1:5000 in Tris-EDTA buffer (10 mM Tris-HCl, 1 mM EDTA, pH 8.0). This dilution is stable for up to 6 h at room temperature when protected from light. Ethidium bromide at 20 µg/L could be used to stain DNA, but it must be handled with care and adequate personal protection equipment must be worn as it is a known mutagen.

10.3.3 Procedure

The illustrated protocol shown in Figure 10.3 is the most commonly used. Nevertheless, when a new cell type and/or a new species is used, lysis step, DNA unwinding, and electrophoresis settings should be optimized. Indeed, agarose concentrations, alkaline unwinding time, electrophoresis time, voltage, and current impact the accuracy of the assay and, therefore, the determination of DNA damage [7].

10.3.3.1 Cell Suspension and Lysis

When studying tissues, such as mussel gills or fish liver, cell suspensions must be obtained either mechanically (mincing tissue in buffer with a pair of scissors or through a mesh membrane) or chemically (by incubation with proteases such as collagenase or trypsin). A low background level of DNA damage reflects a cell suspension of high quality. The choice of the agarose concentration is a function of the average size of DNA fragments to be separated during electrophoresis. NMA concentration is generally 1% but sometimes may be lowered to 0.8%. Generally, lysis runs for at least 1 h. When studying spermatozoa and plant cell DNA damage, a different lysis step is required. For sperm, due to the high compaction of the chromatin in the nuclei, lysis duration time usually has to be significantly increased up to 24 h at 4°C [5]. Genomic DNA decondensation could also be done by nuclei incubation with proteolytic enzymes. For instance, dithiothreitol, proteinase K, or RNase could also be added to the lysis solution [8]. For plant cells, due to the presence of cell wall and plasma membrane, nuclei are directly isolated by mincing roots or leaves with a razor blade on ice [9].

10.3.3.2 DNA Unwinding

The longer the exposure to alkaline buffer, the greater the expression of alkali-labile damage. Generally slides are put in the alkaline buffer at 4°C for 20 min minimum, but time can be increased up to 40 min to allow the expression of additional alkali-labile sites.

Preparation of precoated slides

Coat microscope slides with 1% NMA: dip slides vertically in 1% NMA (in a beaker or in a coplin jar) while agarose is hot (± 55°C) and gently remove. Wipe underside of slide to remove agarose.
Let the slides to air dry horizontally overnight.
Store the slides at room temperature until needed (can be stored for year).

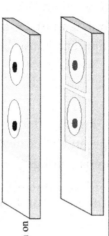

Slide

Hot 1% NMA

Cell isolation and viability check

Almost any eukaryotic cells obtained as a single cell or nuclear suspension could be used.
Cell density should be 1.10^6 cells/mL.
Cell suspensions exhibiting <90% viability should be discarded.

Preparation of Comet slides

Mix same volume of 1% LMPA and of cell suspension (1.10^6 cells/mL). Spread 2 drops of 40 μL each on precoated slide and add 22x22 coverslips. Cell density should be 0.5 to 3.10^4 cells/gel.
Allow gel to polymerise at 4°C on a slide tray resting on ice (5 to 10 minutes).

Eventually a third layer of 0.5% LMPA can be added (40 μl 0.5% LMPA per gel) in the same manner as previously.

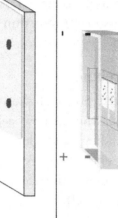

Lysis

From this step, work under dim yellow lights to prevent DNA damage.
Gently slide off coverslips with the thumb on the side of the microscope slide.
Slowly immerse (either vertically in a Coplin jar or horizontally in a tray) slides into cold, freshly made lysing solution for 1 hour at 4°C.

Unwinding

Gently remove slides from the lysing solution and drain.
Place slides side by side on the horizontal gel chamber.
Immerse slides in freshly prepared electrophoresis buffer pH>13 for 20 to 40 min in the dark at 4°C.
Slides must be completely covered by the buffer.

Figure 10.3 Illustrated protocol of the alkaline Comet assay as commonly used in ecotoxicology.

Electrophoresis

Turn on power supply to 0.6-0.8 V/cm.
Perform electrophoresis at 4°C for 20 min under 300 mA.
Adjust the current to 300 mA by raising or lowering the buffer level.

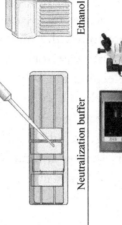

Neutralization

Lift the slides from the electrophoresis chamber and place on a staining tray.
Cover the slides with Neutralization buffer, using a pipette: apply 3 washes for 5 minutes each.

From this step, work under normal lights.
After draining, immerse slides in Coplin jar filled with absolute ethanol for 20 min to dehydrate them, then let them air dry.

Neutralization buffer   Ethanol

Staining and comet visualization

Stain slide with 40µL diluted SYBR Green (or another DNA fluorescent dye), place a coverslip over it and visualize immediately under epifluorescence microscope (x 400 magnification).
Score comets through computerized image analysis system or score visually.

Figure 10.3 (Continued)

10.3.3.3 Electrophoresis

In most cases, slide electrophoresis runs for 20 min, under 300 mA, 0.6–0.8 V/cm, but it is the voltage gradient across the agarose gel coupled with duration time of electrophoresis that determines the migration of DNA. The voltage should remain low (less than 1V/cm) to perform a slow migration of DNA to avoid artifactual DNA damage. Keeping the V/cm × electrophoresis time ratio constant will produce approximately the same shape and size of comets, which allow good reproducibility. A spreadsheet has been proposed by Collins et al. [6] to calculate the effective voltage at which the DNA is subjected. When the optimal settings are found, it is easier and more reliable to fill the electrophoresis tank with the same volume of buffer every time rather than changing the volume of buffer to gain 300 mA precisely.

10.3.4 Positive Controls

Several model genotoxicants can be used as positive control to calibrate the protocol of the Comet assay [10]. The calibration should be done using X-ray or gamma ray exposure as a reference to generate known amounts of DNA damage (1 Gy of gamma rays induces 0.29 strand breaks/10^9 Da DNA) [11].

10.3.5 Data Calculation and Interpretation

After staining, comets are counted under an epifluorescence microscope (×400 or ×200 magnification). DNA damage level is measured through the scoring of 50–100 comets per gel. Two kinds of scoring are generally used: visual scoring or computerized scoring methods, both giving comparable results.

10.3.5.1 Visual Scoring

With visual scoring, slides must be blind labeled to avoid any subjective bias. Score randomly 100 comets per gel. Scale the comets to one of five classes according to comet head and tail intensity. Each comet class is assigned a value between 0 and 4. Examples of comets with the corresponding class are shown in Figure 10.4. DNA damage level reported as Total Comet Score in arbitrary units, is calculated as follows: Total Comet Score = $(0 \times n_0 + 1 \times n_1 + 2 \times n_2 + 3 \times n_3 + 4 \times n_4)/(\sum/100)$ where n_0 is the number of comets in class 0, n_1 the number of comets in class 1, ... [12].

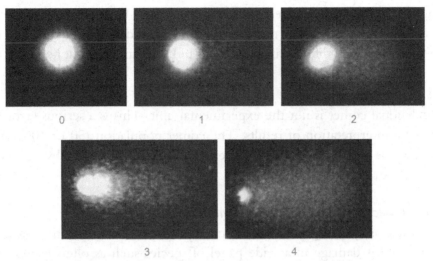

Figure 10.4 Comet images with their corresponding class value for visual scoring [29].

This type of scoring method shows linear correlation with the percentage of DNA in the comet tail measured with image analysis systems: DNA damage (or % Tail DNA) = Total Comet Score/4 [13].

10.3.5.2 Computerized Scoring

There are several available kinds of software to score comets, and some of them are free on the Internet. Image analysis software can calculate up to 30 different parameters for each comet [14]. The user must select the comets randomly and images are analyzed by the software based on the fluorescence intensity. Area of the comet, head and tail intensities, tail length, and tail moment (tail length × relative tail intensity) are calculated. The tail intensity (percentage of DNA in the tail) is the recommended parameter because it bears a linear relationship to break frequency, it is essentially unaffected by threshold settings, and it allows discrimination damage over the widest possible range. It also provides a simple visualization of what comets look like [15]. A few totally automated scoring systems are available, but their use remains atypical. Comets with almost all DNA in the tail and with a very small head, equivalent to "class 4" comets, according to visual scoring and called hedgehogs, are not always dead cells and do not represent apoptotic cells. They should be considered when scoring [16]. Indeed, apoptosis leads to a very thin fragmentation of DNA (of the order of oligonucleotides size). This kind of fragment cannot be detected after electrophoresis. However, to avoid taking into account dead cells, it is recommended to exclude cells looking like halos (comet "ghosts" with diffuse small fragments of DNA and without heads).

10.3.5.3 Statistical Analysis

The individual comet is not the experimental unit. This is a serious error leading to false interpretation of results. The comet population (50 or 100 comets) represents one experimental data. Generally, these data are given by the median of comets scored.

10.3.5.4 Applications in Ecotoxicology Interpretation

The Comet assay has gained more and more attention in ecotoxicology for detecting DNA damage in a wide panel of species such as oligochaetes, polychaetes, planarians, cnidarians, insects, crustaceans, bivalves, gastropods, asteroids, echinoids, fishes, amphibians, and mammals [17]. It is extensively used both *in vitro* and *in vivo* to study DNA damage mechanisms and to evaluate the genotoxic potential of chemicals, sediments, air particles, and effluents (wastewaters). A protocol for cell lines, initially used in toxicology, has been extended with success to the field of ecotoxicology and offers the possibility to study DNA damage and repair using nondestructive means by taking a blood/hemolymph/excreted liquids sample, thus respecting the "3 R" concept—reduction, refinement, replacement—in toxicology investigation.

The most commonly used cell types are erythrocytes and hemocytes, due to their ease of collection, and hepatocytes and gill cells because of their metabolic and physiologic functions. The use of spermatozoa is also particularly relevant because of their involvement in reproductive function [8,18,19]. The Comet assay can be efficiently combined with the use of other biomarkers for answering complex environmental questions in the field [17,20]. With the *in situ* approach, it is possible to explain the natural variability of biological responses faced with genotoxic stress, such as the influence of abiotic factors: temperature, calcium—carbon balance, or photoperiod [18,21,22] and reproductive status, age, or gender [23].

Strand breaks detected by the Comet assay can be due directly to genotoxic agents, but frequently, they are transient strand breaks issued from the DNA repair process activity or alkali-labile sites are converted to strand breaks under alkaline conditions. The relevancy of such transient damage in terms of effects is low. Furthermore, a high level of strand breaks indicates either high DNA damage or efficient repair capacity. To get further useful information about the type of damage, several modified protocols have been applied. The use of restriction endonucleases allows us to detect more specifically some types of DNA damage and to increase the sensitivity of the assay. In brief, after lysis, slides are neutralized in adequate enzyme buffer. Before unwinding and electrophoresis, nucleoids are incubated for 30–60 min at 37°C with specific enzymes: endonuclease III (specific of oxidized pyrimidines), formamidopyrimidine glycosylase (specific of ring-opened purine, purine oxidation products, and N-alkylation products), 3-methyladenine DNA glycosylase II (specific of O- and N-alkylation products), or T4 endonuclease V (specific of cyclobutane pyrimidine dimers) [15,24]. The enzyme digestion step leads to an increase in the DNA migration related to the damage type targeted by the chosen enzyme. Such modified protocols offer interesting opportunities to study base excision repair and nucleotide excision repair activities [23,25]. Combined with fluorescence *in situ* hybridization (Comet-FISH), the Comet assay allows us to reveal region-specific DNA damage and repair genes or sequences and to get further insights into the sensitivity of specific DNA regions [26].

10.3.6 Advantages and Disadvantages
10.3.6.1 Advantages

- The Comet assay requires a very small sample of any eukaryotic nucleated cell (less than 10,000 cells) and it can be further miniaturized to decrease cell number or increase its throughput.
- It is less expensive and more sensitive than several genotoxicity tests such as chromosomal aberration, sister chromatid exchange, and alkaline elution.
- The scoring is based on individual cells, allowing the study of the distribution of DNA damage in cell populations.
- Equipment required is easy to obtain and relatively cheap: epifluorescence microscope, agarose gel apparatus, and a power supply. Radioactive labeling is not required.
- The use of specific endonucleases may improve the sensitivity of the assay and allows the detection of specific DNA lesions such as FormamidoPyrimidine Glycosilase (Fpg) for opened-ring purine, N-alkylation products, Endonuclease III for oxidized pyrimidines, and T4

endonuclease V reveals specific cyclobutane pyrimidine dimers, an UV-induced damage type.

- DNA repair capacity can be studied easily by kinetic analysis of the recovery of DNA damaged cells.
- It has been recommended by many researchers for regulatory studies to replace traditional tests such as the OECD unscheduled DNA synthesis (UDS) test with liver cells [27,28].

10.3.6.2 Limitations

- Obtaining a cell suspension of high quality may be tricky when working with tissues.
- The major disadvantage is the lack of standardization of this method, resulting in difficulties when comparing the results from different laboratories.
- Comet scoring may be user dependent and subjective if slides are not blindly marked. Scoring methods, either visual or semi-automated, should be calibrated internally in the laboratory for each cell type studied.

Several key articles have been written concerning the protocol of the Comet assay and its applications in ecotoxicology. A good review, clearly organized as tables, presents the use of the Comet assay in different models from bacteria to man, employing various cell types [29]. Another review demonstrates the ecotoxicological relevance of the assay, highlighting the correlation between the Comet response with other relevant biological responses and biomarkers [30]. Although written for toxicologists rather than ecotoxicologists, the article written by Azqueta and Collins broadens the scope of use of the Comet assay and the modifications that can be applied to detect various specific DNA damage [31]. A clear presentation of the Comet assay focusing on aquatic genotoxicology gives new prospects of use [24].

10.4 DNA CONTENT VARIATION IN CELLS

10.4.1 Introduction

Nuclei that undergo cytogenetic damage, such as loss of chromosome parts or formation of micronuclei, could be determined by measuring the variation of DNA content in cells by flow cytometry. A flow cytometer is a cell- or particle-based light scattering and fluorescence instrument that can quantitatively determine these properties on a cell-by-cell basis. The principle of the assay is simple: cells are fixed and treated to RNase A to degrade RNA, and the cells are then DNA stained by dyes (SYTO Green or propidium iodide dyes) consistent with the excitation laser of the flow cytometer (argon lasers emit at 488 nm or helium-neon lasers at 632 nm). The stained cells are then analyzed by a flow cytometer where the number of cells and DNA fluorescence values are counted by the instrument. The coefficient of variation of the mean DNA content of 2 N (resting cells) is indicative of DNA cytogenetic damage [32]. This analysis could also identify aneuploid (1 n) or tetraploid (4 n) cells, which could be indicative of transformed cells. However, tetraploid cells also are the result of dividing cells where the number of DNA was duplicated before cell division and care should be taken when analyzing the data. For example, dividing tetraploid cells should give some octoploid cells as well.

10.4.2 Reagents

Cell fixative: Add formaldehyde to give 3% final concentration in PBS containing 1 mM EDTA. PBS is composed of 145 mM NaCl, 5 mM KH_2PO_4, and 1 mM $NaHCO_3$, and the pH is adjusted at 7.4 with HCl 0.1 N. For marine invertebrate cells, 1−1.5 g of NaCl could be added for 100 mL of PBS to ensure osmolarity.

DNA staining solution: Prepare 20 µg/mL of propidium iodide and 20 U/mL RNase A in 50−100 mL of PBS containing 1 mM EDTA. Concentrate solution ($10-20 \times$) could be separately prepared in PBS for RNase A (store at $-20°C$) and propidium iodide (store at $-20°C$) to prepare the previous working solution (prepared daily).

10.4.3 Procedure and Data Collection

A volume of 250 µL of cell suspension (isolated erythrocytes, leukocytes, cell lines, or hemocytes) is mixed with 750 µL of cell fixative solution at 4°C for 30 min. The cells are then centrifuged at $1000 \times g$ for 1 min and the supernatant removed by aspiration. The cells are then resuspended in DNA staining solution and allowed to incubate for 60 min at 37°C to allow for RNA digestion.

Using a typical argon laser flow cytometer (Becton Dickinson Company), cells were first gated from debris (devoid of DNA) by gating cells on a 2D plot: cell volume (orthogonal light scatter) and cell complexity (side light scatter). Between 10,000 and 20,000 cells are collected by the instrument and a second histogram (as shown below) is constructed with the cell events number and DNA content on the *y*- and *x*-axis, respectively. The mean with SD is

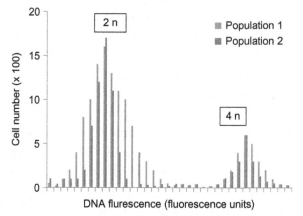

Figure 10.5 DNA content in two hypothetic cell populations. The DNA fluorescence values for each particle are determined in a given homogenous cell population by flow cytometry. The mean DNA value with its SD is then calculated for populations 1 and 2. The SD of DNA content for 2n cells represents the metric of genomic DNA integrity. The variation in DNA (SD) of population 1 is larger than population 2.

then calculated by the instrument software for the 2n cells for populations 1 and 2. The SD for DNA content represents a measure of DNA integrity. In Figure 10.5, both cell populations exhibit the same mean DNA content but the variation of population 2 is greater than population 1. DNA content in cells from population 2 is more variable hence indicative of cytogenetic damage. To visualize this on a larger scale, damaged cells undergoing division will produce micronucleated cells, therefore, producing one cell with more DNA (nuclei + micronuclei) and cells with less DNA (nuclei − micronuclei). The mean content would be the same, but the variation in DNA content would be larger is such cases. The mitotic index could also be estimated by the number of 4n cells/the number of 2n cells. This parameter is of interest because, as explained below, cell division rate will influence the number of cells with variable DNA content.

10.5 MICRONUCLEI

The evaluation of micronucleated cells is a simple and cheap way to assess the occurrence of DNA cytogenetic damage in cells. It is simply based on the direct observation of micronucleated cells on a microscope at $200-400 \times$ magnification. Micronucleated cells are produced during cell division where an

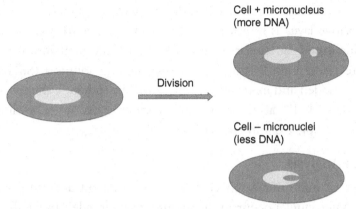

Cell + micronucleus
(more DNA)

Division

Cell – micronuclei
(less DNA)

Figure 10.6 The process of micronuclei formation in cells.

extra amount of DNA occurs as micronuclei orbiting the nuclei (Figure 10.6). However, since the method is based on observation, it is considered semiquantitative and care should be taken to maintain the person's objectivity such as "blinding" the slide identification (by numbers) during the cell scoring procedure.

10.5.1 Reagents

Cell fixative: Add formaldehyde to give 3% final concentration in PBS containing 1 mM $MgCl_2$. PBS is composed of 145 mM NaCl, 5 mM KH_2PO_4, and 1 mM $NaHCO_3$, and the pH is adjusted at 7.4 with HCl 0.1 N. For marine invertebrate cells, 1–1.5 g of NaCl could be added for 100 mL of PBS to ensure osmolarity.

DNA staining solution: Giemsa staining or H&E staining solutions are commercially available. Making your own stain could be tedious and many are commercially available. Dissolve 1 g Giemsa dye powder (molecular formula $C_{14}H_{14}ClN_3S$) in 66 mL of glycerol at 60°C for 2 h. Bring to 37°C and add 66 mL of methanol and shake until maximum dissolution of the dye. Paper filter (Whatman #4) the solution, followed by a filtration on a 0.45 μm pore filter. Store at room temperature in a dark bottle.

10.5.2 Procedure

A 100–200 μL volume of the cell suspension (e.g., fish leukocytes and bivalve hemocytes) is smeared on poly-L-lysine microscope glass slides and incubated for 1 h in saturated humidity to allow for cell adherence. The supernatant is removed and 300 μL of fixative solution is added dropwise to completely cover the smeared area. The fixative layer is removed after 10 min and 300 μL of 90% methanol is added and incubated on ice for 30 min. Afterward, the methanol is removed and the slides are dried and 10 drops of Giemsa staining solution (Sigma Chemical Company) are added and incubated for 30 min. The slides are then washed in distilled water or with 1% acetic acid if heavy staining occurs. The slides are then ready for microscopic examination at 200−400 × enlargement.

10.5.3 Data Analysis

The number of micronucleated cells is determined on a 1000 cells basis (per 1000, $^o/_{oo}$). Micronuclei occur in the vicinity of the nuclei. In parallel, the number of tetraploid cells (cells undergoing division) was also counted to determine the relative cell division rate, which is $\sim 0.01-0.2\%$ in aquatic organisms.

10.6 DNA SYNTHESIS AND CATABOLISM

10.6.1 Dihydrofolate Reductase

Dihydrofolate reductase (DHFR) is involved in the reduction of dihydrofolate to tetrahyfolate, which is a precursor for purine synthesis in cells:

$$\text{Dihydrofolic acid} + \text{NADPH} + \text{H}^+ \rightarrow \text{Tetrahydrofolic acid} + \text{NADP}^+$$

This enzyme is found in all cells and represents an important enzyme for purine production. It is noteworthy that some antibiotics and antineoplasic agents (methotrexate) are potent inhibitors of DHFR activity in cells, hence their use in therapeutics.

10.6.1.1 Reagents

Assay buffer: Prepare 10 mM Tris−acetate, pH 7.2, containing 2 mM MgCl$_2$ and 1 mM dithiothreitol.

Reduced NADPH: Prepare 1 mM stock in 20 mM NaOH. Store in aliquots at $-80°$C.

Dihydrofolate substrate: Prepare a 10 mM master solution in 50 mM KH$_2$PO$_4$, pH 7.2. Dilute ½ with the assay buffer just before the assay.

10.6.1.2 Procedure

In a black microplate, prepare the following in duplicates: 50 μL tissue homogenate extract or assay buffer (blank), 110 μL of assay buffer, 20 μL of reduced

NADPH (or water for the blank), and 20 μL dihydrofolate substrate. Incubate at 0, 10, 20, and 40 min and determine NADPH by excitation at 350 nm and emission at 460 nm. The rate of NADPH disappearance is determined.

Controls could be included to identify potential artifacts. A negative control consists of adding the S12 extract in the presence of NADPH to determine the background oxidation of NADPH. The rate of disappearance of NADPH should be greater in the presence of dihydrofolate substrate than without the substrate. Samples with strong NADPH oxidase activity could represent a false positive sample.

10.6.1.3 Data Calculation and Analysis

DHFR activity is determined by the following:

$$\text{NADPH fluorescence } t0 - t30 \text{ min} \times 1/30 \text{ min} \times 1/0.05 \text{ mL}$$
$$= \text{rate of NADPH}/(\text{min} \times \text{mL})$$

The enzyme activity is then normalized against total proteins (mg/mL) or tissue weight (g/mL) to give: decrease NADPH/(min × mg proteins or g tissue weight).

10.6.2 Aspartate Transcarbamoylase Activity

This enzyme catalyzes the condensation of carbamoyl phosphate with aspartate to form carbamyl aspartate, which is a precursor for pyrimidine bases (cytosine, thymidine, and uracil) during DNA synthesis. The principal of the assay is based on the release of phosphate during the reaction:

$$\text{Aspartate} + \text{carbamoyl phosphate} \rightarrow \text{carbamoyl aspartate} + \text{Pi}$$

The assay separates the phosphomolybdate complex from molybdate, which can release phosphate from carbamoyl phosphate in acidic conditions during the phosphate assay. Phosphate is determined colorimetrically using the phosphomolybdate assay [33] and carbamoyl phosphate is removed using the citrate/butanol extraction method developed by Herries [34].

10.6.2.1 Reagents

Assay buffer: Prepare 25 mM HEPES buffer, adjust pH to 7.
Aspartate: Prepare 1% solution in water (100 mL).
Carbamoyl phosphate: Prepare 1 mM solution in 100 mL of bidistilled water (150 mg/100 mL).

Molybdate reagent: Dissolve 1 g ammonium molybdate in 100 mL of 0.9 M H_2SO_4.

Butanol: Analytical grade 100%.

Citrate: Prepare 1% solution by dissolving 1 g citrate (trisodium salt) in 100 mL water.

Phosphomolybdic acid stock: 10% phosphomolybdic acid in ethanol.

Blank solution: In a 50 mL tube, add 4 mL of aspartate, 4 mL of carbamoyl phosphate, and 4 mL of assay buffer. Mix and add 3 mL of molybdate and 12 mL of butanol and mix well and add 6 mL of citrate. Mix thoroughly with a vortex mixer and let the phases separate. Collect the butanol phase and transfer into another 50 mL tube.

Phosphomolybdate standard: A stock solution of phosphomolybdate 10% in ethanol is prepared (Sigma). Dilute with the blank solution to obtain a final concentration 0.25%.

10.6.2.2 Procedure

Prepare four microcentrifuge tubes denoted as follows: (1) reaction media, (2) stopped reaction $t = 0$, (3) stopped reaction $t = 15$ min, and (4) stopped reaction $t = 30$ min.

In tube 1, take 200 μL of S12 and mix with 400 μL aspartate, 400 μL carbamoyl phosphate, and 400 μL assay buffer.

In tubes 2, 3, and 4, take 400 μL of reaction mixture (tube 1) at the corresponding times and add 100 μL molybdate, 400 μL butanol (mix), and 200 μL of citrate (mix). Briefly centrifuge at $10,000 \times g$ for 20–30 s to separate the phases.

In a clear microplate, transfer 150 μL of butanol phases in duplicates (tubes 2, 3, and 4). Add 10 μL of methanol and read absorbance at 340 nm. A standard well (43 μM) is prepared by diluting the phosphomolybdate standard (5 μL) with the stock blank solution (145 μL) in duplicates. The blank is 150 μL of blank stock solution.

10.6.2.3 Data Analysis and Calculation

[(A340 sample - Ablank)30 min-(A340 sample - Ablank)0 min]

   $\times$ 1/30 min $\times$ [43 nmol/mL/(A340 - Ablank] $\times$ dilution factor

   (1.4/0.2) = nmol released phosphate/min.mL

The activity is then normalized against total proteins (mg/mL) or homogenate weight (g/mL) to give the final activity expressed as nmol of released phosphate/min/mg proteins or g tissue.

10.6.3 Xanthine Oxidoreductase (XOR)

Xanthine oxidoreductase (XOR) is involved in the degradation of purines in organisms. Increased degradation of purines could lead to oxidative stress since the reaction involves the release of reactive oxygen species (hydrogen peroxide). XOR catalyzes the addition of one oxygen atom to hypoxanthine to give xanthine:

$$\text{hypoxanthine} + H_2O + O_2 \rightarrow \text{xanthine} + H_2O_2$$

Xanthine can be further oxidized into uric acid (and the formation of additional H_2O_2) in cells but it remains unclear at present if this activity is present in invertebrates. During this reaction, hydrogen peroxide is released, which can be conveniently quantified by fluorometry using the dichlorofluorescein/peroxidase system.

Xanthine oxidase is a form of XOR, a type of enzyme that generates reactive oxygen species. These enzymes catalyze the oxidation of hypoxanthine to xanthine and can further catalyze the oxidation of xanthine to uric acid.

10.6.3.1 Reagents

Assay buffer: Prepare 50 mM KH_2PO_4 buffer containing 10 μM ammonium molybdate. Adjust pH to 7.4.

Hypoxanthine: Prepare a 500 mM stock solution in the assay buffer.

Dichlorofluorescein (DCF): Prepare a 10 mM stock solution in DMSO 100%.

Peroxidase: Dissolve horseradish peroxidase in 50 mM KH_2PO_4 buffer, pH 6, to obtain 5 mg/mL. Add DCF 10× to obtain 100 μM final concentration.

DCF/peroxidase reagent: Mix 100 μL of DCF with 20 μL of peroxidase and complete to 100 mL with assay buffer.

10.6.3.2 Procedure

In an opaque microplate for fluorescence, transfer 50 μL of S12 (12,000 × *g* supernatant) fraction of the homogenate or blank (water). Add 125 μL of assay buffer, 5 μL of hypoxanthine, and 20 μL of DCF/peroxidase reagent. Incubate at 30°C for 0, 10, 20, and 40 min and measure the formation of fluorescein at 485 nm excitation and 535 nm emission. In this assay, different blanks could be prepared to understand the background effects of the tissue extracts from various organisms and tissues:

Blank 1: 50 μL of water or homogenization buffer, 125 μL of assay buffer, 5 μL of hypoxanthine, and 20 μL of DCF/peroxidase reagent (intrinsic oxidation of hypoxanthine)

Blank 2: 50 μL of sample, 125 μL of assay buffer, 5 μL of water, and 20 μL of DCF/peroxidase reagent (basal hypoxanthine or H_2O_2 levels in the biological sample)

Blank 3: 50 μL of sample, 125 μL of assay buffer, 5 μL of water, and 20 μL of water (autofluorescence of the biological sample)

Calibration is achieved by including a 200 μL blank (assay buffer) and 200 μL of 0.5—1 μM fluorescein standard. Read fluorescence at the same wavelengths above.

10.6.3.3 Data Calculation and Handling

The enzyme activity is calculated by the following :

$$[(\text{Fluorescence sample} - \text{Fluorescence blank 2})_{40\ min}$$
$$- ([\text{Fluorescence sample} - \text{Fluorescence blank 2})_{0\ min}$$

and this is multiplied with the standard fluorescence as follows:

$$[(\text{Fluorescence concentration (nmol/mL)}]$$
$$/[\text{Fluorescence standard} - \text{Fluorescence assay buffer}]$$

to give Fluorescein formed/mL. This concentration is then multiplied with the incubation time (1/40 min) and the dilution factor (200/50) to give the XOR activity nmole fluorescein formed/min/mL.

This activity is then normalized by total protein content (mg/mL) of the S12 fraction or homogenate weight (g/mL) to give nmole fluorescein formed/min/mg proteins. With samples of low activity, the enzyme activity could be calculated using an initial time in the linear portion of fluorescein formation in the kinetic measurement. For example, fluorescence readings at $t = 10$ and $t = 40$ min could be used provided the increase is linear in time.

REFERENCES

[1] Olive PL. DNA precipitation assay: a rapid and simple method for detecting DNA damage in mammalian cells. Environ Mol Mutagen 1988;11:487−95.

[2] Gagné F, Blaise C. Genotoxicity of environmental contaminants in sediments to rainbow trout hepatocytes. Environ Toxicol Wat Qual 1995;10:217−29.

[3] Bester MJ, Potgieter HC, Vermaak WJH. Cholate and pH reduce interference by SDS in the determination of DNA with Hoechst. Anal Biochem 1994;223:299−305.

[4] Singh NP, McCoy MT, Tice RR, Schneider EL. A simple technique for quantitation of low levels of DNA damage in individual cells. Exp Cell Res 1988;175:184−91.

[5] Lacaze E, Geffard O, Bony S, Devaux A. Genotoxicity assessment in the amphipod Gammarus fossarum by use of the alkaline comet assay. Mutat Res 2010;700:32−8.

[6] Collins AR, Oscoz AA, Brunborg G, Galvao I, Giovannelli L, Kruszewski M, et al. The comet assay: topical issues. Mutagenesis 2008;23:143−51.

[7] Azqueta A, Gutzkow KB, Brunborg G, Collins AR. Towards a more reliable comet assay: optimising agarose concentration, unwinding time and electrophoresis conditions. Mutat Res 2011;724(1-2):41−5.

[8] Baumgartner A, Cemeli E, Anderson D. The comet assay in male reproductive toxicology. Cell Biol Toxicol 2009;25(1):81−98.

[9] Ventura L, Giovannini A, Savio M, Dona M, Macovel A, Buttafava A, et al. Single cell gel electrophoresis (comet) assay with plants: research on DNA repair and ecogenotoxicity testing. Chemosphere 2013;92(1):1−9.

[10] Kirkland D, Kasper P, Müller L, Convi R, Speit G. Recommended lists of genotoxic and non-genotoxic chemicals for assessment of the performance of new or improved genotoxicity tests: a follow-up to an ECVAM workshop. Mutat Res 2008;653(1−2):99−108.

[11] Johansson C, Møller P, Forchhammer L, Ofts S, Godschalk RW, Langie SA, et al. An ECVAG trial on assessment of oxidative damage to DNA measured by the comet assay. Mutagenesis 2010;25(2):125−32.

[12] Azqueta A, Shaposhnikov S, Collins AR. DNA oxidation: investigating its key role in environmental mutagenesis with the comet assay. Mutat Res 2009;674:101−8.

[13] Garcia O, Romero I, González JE, Moreno DL, Cuetara E, Rivero Y, et al. Visual estimation of the percentage of DNA in the tail in the comet assay: evaluation of different approaches in an intercomparison exercise. Mutat Res 2011;720:14−21.

[14] Kumaravel TS, Vilhar B, Faux SP, Jha AN. Comet assay measurements: a perspective. Cell Biol Toxicol 2009;25:53−64.

[15] Collins AR. The comet assay for DNA damage and repair. Mol Biotechnol 2004;26:249−61.

[16] Lorenzo Y, Costa S, Collins AR, Azqueta A. The comet assay, DNA damage, DNA repair and cytotoxicity: hedgehogs are not always dead. Mutagenesis 2013;28(4):427−32.

[17] Jha AN. Ecotoxicological applications and significance of the comet assay. Mutagenesis 2008;23:207−21.

[18] Lacaze E, Devaux A, Jubeaux G, Mons R, Gardette M, Bony S, et al. DNA damage in Gammarus fossarum sperm as a biomarker of genotoxic pressure: intrinsic variability and reference level. Sci Total Environ 2011;409:3230−6.

[19] Santos R, Palos-Ladeiro M, Besnard A, et al. Relationship between DNA damage in sperm after ex vivo exposure and abnormal embryo development in the progeny of the three-spined stickleback. Reprod Toxicol 2013;36:6−11.

[20] Frenzilli G, Nigro M, Lyons BP. The comet assay for the evaluation of genotoxic impact in aquatic environments. Mutat Res 2009;681:80−92.

[21] Buschini A, Carboni P, Martino A, Poli P, Rossi C. Effects of temperature on baseline and genotoxicant-induced DNA damage in haemocytes of Dreissena polymorpha. Mutat Res 2003;537(1):81−92.

[22] Rank J, Lehtonen KK, Strand J, et al. DNA damage, acetylcholinesterase activity and lysosomal stability in native and transplanted mussels (Mytilus edulis) in areas close to coastal chemical dumping sites in Denmark. Aquat Toxicol 2007;84(1):50−61.

[23] Akcha F, Vincent HF, Pfhol-Leszkowicz A. Potential value of the comet assay and DNA adduct measurement in dab (*Limanda limanda*) for assessment of in situ exposure to genotoxic compounds. Mutat Res 2003;534:21—32.

[24] Devaux A, Bony S. Genotoxicity of Contaminants: Comet Assay. In: Férard J-F, Blaise C, editors. Encyclopedia of Aquatic Ecotoxicology. Dordrecht, The Netherlands: Springer; 2013.

[25] Kienzler A, Bony S, Devaux A. DNA repair activity in fish and interest in ecotoxicology: a review. Aquat Toxicol 2013;134—135:47—56.

[26] Glei M, Hovhannisyan G, Pool-Zobel BL. Use of Comet-FISH in the study of DNA damage and repair: review. Mutat Res 2009;681:33—43.

[27] Tice RR, Agurell E, Anderson D, Burlinson B, Hartmann A, Kobayashi H, et al. Single cell gel/comet assay: guidelines for in vitro and in vivo genetic toxicology testing. Environ Mol Mutagen 2000;35:206—21.

[28] Hartmann A, Plappert U, Poetter F, Suter W., et al. Comparative study with the alkaline Comet assay and the chromosome aberration test. Mutat Res 536(1—2):27—38.

[29] Dhawan A, Baijpayee M, Parmar D. Comet assay: a reliable tool for the assessment of DNA damage in different models. Cell Biol Toxicol 2009;25:5—32.

[30] Jha AN. Genotoxicological studies in aquatic organisms: an overview. Mutat Res 2004;552:1—17.

[31] Azqueta A, Collins AR. The essential comet assay: a comprehensive guide to measuring DNA damage and repair. Arch Toxicol 2013;87:949—68.

[32] Debenest T, Gagné F, Burgeot T, Blaise C, Pellerin J. DNA integrity assessment in hemocytes of soft-shell clams (*Mya arenaria*) in the Saguenay Fjord (Québec, Canada). Environ Sci Pollut Res Int 2013;20:621—9.

[33] Stanton MG. Colorimetric determination of inorganic phosphate in the presence of biological material and adenosine triphosphate. Anal Biochem 1968;22:27—34.

[34] Herries DG. The simultaneous estimation of orthophosphate and carbamoylphosphate and application tho the aspartate transcarbamoylase reaction. Biochem Biophys Acta 1967;136:95—8.

[35] Houtgraaf JH, Versmissen J, van der Giessen WJ. A concise review of DNA damage checkpoints and repair in mammalian cells. Cardiovasc Revasc Med 2006;7:165—72.

CHAPTER 11

Biomarkers of Infection and Diseases

François Gagné

Chapter Outline

The purpose of this chapter is to provide insights on some methods for determining the state of infection of organisms during ecotoxicology studies. In the environment, organisms have to cope with various physical (temperature, light), chemical (pollutants, nutrients), and biological (microorganisms) variables. Indeed, the organisms are seldom exposed to only one stressor at a time. Organisms constantly have to defend themselves against invading foreign organisms in addition to exposure to contaminants. Measuring marker enzymes of exposure to bacteria (e.g., β-galactosidase or peptidases), viruses (e.g., reverse transcriptase activity, neuraminidase), and some cyanobacteria (inhibition of protein phosphatase) are beyond the scope of this book, but some guidance will be provided for the reader. However, the organism's defense systems against

197

exposure and deleterious effects to microorganisms will be provided. The immune system varies in complexity across the phyla, ranging from the production of toxins by unicellular organisms to the highly developed immune system in vertebrates. Vertebrate immune systems comprise both cellular (nonspecific) and acquired (antibodies) immunity by a complex array of mediators in white blood cells (leukocytes) composed of lymphocytes, granulocytes, and monocytes. The vertebrate immune system possesses the capacity to learn and acquire immunity by the production of antibodies when exposed initially to the pathogens. The innate immunity is composed of both cellular (phagocytosis) and humoral mechanisms, which involve the production of cytokines for cell killing (natural cell cytotoxicity, nitric oxide production, lysozyme activity, etc.). Invertebrates rely mostly on innate immunity and do not possess the complement system for acquired immunity functions such as the production of antibodies. Despite this, invertebrates are able to thrive in environments that abound in microorganisms such as bacteria-rich sediments and surface waters.

Although some commonly used biochemical markers are proposed in this chapter to enable the investigators to gain some insights about the health status of organisms in respect to infections, they are not intended to be comprehensive in any way. Indeed, there exist a number of ways to characterize the immune systems in an organism, and the reader is invited to consult the following reviews [1−4]. The biochemical assays presented here were successfully used in pilot field studies because of their ease of analysis and convenience. Organisms, especially invertebrates, produce a number of humoral factors to limit the infestation of microorganisms. The release of lysozymes in the hemolymph or sera of organisms could represent a simple way to determine the proliferation of bacteria in an organism [5]. The presence of microorganisms could also lead to inflammation by increased arachidonate-dependent cyclooxygenase in tissues and increased production of nitric oxide (NO), which follows oxidative burst responses in immune cells to destroy ingested foreign bodies from phagocytosis [6]. Ingested microorganisms are destroyed by a series of reactions that fall under oxidative burst. During that process, NO is produced with hydrogen peroxide to form the highly reactive oxygen product, peroxynitrite, leading to the destruction of ingested bacteria and viruses. NO is produced by NO synthase activity and sustained activity could lead to the formation of NO adducts by proteins and DNA, which could induce oxidative-mediated damage to the host cell.

11.1 BIOMARKERS OF EXPOSURE

Exposure of organisms to microorganisms consists of measuring specific biochemical aspects of the invading pathogen and is not found in the host or found at low levels. For example, β-galactosidase activity is readily present from certain bacteria where the (eukaryote) host does not express the enzyme. Hence, the increased presence of β-galactosidase in tissues (e.g., gills, digestive gland, or liver) is indicative of the presence of bacteria in tissues. An overview of biochemical markers of exposure to microorganisms is presented in Table 11.1.

The markers cited in Table 11.1 are biochemical properties specific to the target microorganism, and during tissue invasion their activity or presence will increase above the background levels. In this sense, they are considered biomarkers of exposure to the invading agents but do not provide information on the pathophysiological effects to the host. These are such biomarkers, i.e., biomarkers of exposure to infectious agents.

Table 11.1 A Selection of Genomic and Enzyme Markers for Microorganisms

Species	Markers	References
Bacteria	Total heterotrophic counts	[7]
	β-galactosidase	[7]
	β-glucuronidase	[8]
	Multi-enzyme profiling	
Cyanobacteria	Phosphatase inhibition	[9]
Microcystis sp. (hepatotoxins)	Protein phosphates	[9]
Alexandrium sp. (neurotoxins)	Microcystin-LR	[10]
	Bromoperoxidase	[11]
	Saxitoxins	[12]
Virus	Neuraminidase, specific gene sequence	[13,14]
General	(qPCR)	[15]
RNA virus	Reverse transcriptase	[16]
	Hyaluronidase	
Parasites (e.g., Giardia, cryptosporidium)	Specific antibodies (immunoassays)	[17]
	Gene (specific) detection by qPCR	[18]

Note: qPCR, quantitative polymerase chain reaction.

11.2 LYSOZYME ACTIVITY

Lysozymes are a family of hydrolytic enzymes produced by the host that degrade the bacterial cell wall. They catalyze the hydrolysis of the 1,4-β link between N-acetylmuramic acid and N-acetyl-D-glucosamine of the bacterial cell wall. Lysozymes abound in egg white and are found in tears, saliva, mucus, and plasma (hemolymph). The assay presented here is based on the microplate-based hydrolysis of inactivated bacterial suspension of *Micrococcus lysodeikticus* [19]; if no microplate reader is available, then use any spectrometer that can measure absorbance at 450 nm (turbidity). It is therefore a cheap, rapid, and convenient assay that can be easily performed in virtually any laboratory. The assay is applicable to any organisms or tissues suspected of being infected by bacteria. It is usually performed on plasma or hemolymph with or without immune cells or in tissues exposed to the environment such as gills, digestive gland/liver, or skin.

11.2.1 Reagent

Assay buffer: Prepare a 100 mM KH_2PO_4 at pH 6.2.

M. lysodeikticus: Prepare a suspension of 0.4 mg/mL of inactivated *M. lysodeikticus* (Sigma Chemical Company) in the assay buffer. The suspension forms a turbid solution (which is normal). The suspensions could be stored in 200 μL aliquots at $-20°C$.

11.2.2 Procedure

To 50−100 μL plasma or hemolymph or S10 fraction of tissue homogenates (e.g., gills or the digestive gland), add 50 μL of *M. lysodeikticus* suspension and 100 μL of assay buffer. Incubate at 25°C for 0, 10, 20, and 30 min and read the absorbance at 450 nm. If no microplate reader is available, then a blank (sample replaced by water and the bacterial suspension) and 40 min tube could be prepared by completing to 1 mL with the assay buffer. A decrease in turbidity from the hydrolytic action of lysozyme is expected to be observed and decrease the absorbance at 450 nm.

11.2.3 Data Calculation

Lysozyme activity is determined by the following:

$$\text{hydrolysis activity} = (A450_{t=0} - A450_{t=30\ min})/30\ min \times$$
$$\text{dilution factor } (200\ μL/50\ μL \text{ of test sample}).$$

The enzyme activity is then normalized against total protein concentration in the test sample or g tissue weight.

11.3 NO SYNTHASE ACTIVITY

NO is involved in the formation of peroxynitrite for killing ingested microorganisms following phagocytosis and consists of a pro-inflammatory mediator. NO possesses strong vasodilator properties and acts as a neuromediator during, for example, the learning of avoidance behavior. This enzyme produces NO from arginine in the presence of an electron donor as follows:

$$\text{L-arginine} + \text{NADPH} + \text{H}^+ + 2\text{O}_2 \rightleftharpoons \text{citrulline} + \text{NO} + \text{NADP}^+.$$

The loss of NADPH fluorescence or the appearance of NO could be determined by direct fluorescence measurement or the Griess reagent for nitrites, since NO readily oxidizes into nitrites (NO2^-) and nitrates (NO3^-). Nitrates are converted into nitrites by nitrate reductase in the presence NADPH.

11.3.1 Reagents

Assay buffer: Prepare 25 mM KH_2PO_4 and adjust the pH at 7.4 (500 mL).

Arginine solution: Prepare a 5 mM arginine stock solution in the assay buffer.

Griess reagent: Commercially available as solid (Sigma Chemical Company). If not available the reagent could be prepared by mixing 25 mL of 0.1% N-(1-naphthyl)ethylenediamine dihydrochloride with 25 mL of 1% sulfanilic acid in 5% phosphoric acid.

Nitrite standard: Prepare a 0.1 M $NaNO_2$ concentration in water.

Nitrate standard: Prepare a 0.1 M $NaNO_3$ concentration in water.

Reduced NADPH: Dissolve 25 mg NADPH in 3.42 mL in 0.02 M NaOH. Conserve in 100 µL aliquots at $-85°C$, which is stable for 1 month. At the day of the assay, dilute 1/5 with the assay buffer.

Nitrate reductase: Dissolve the lyophilized powder in water to obtain 6.4 U/mL. Conserve in 100 µL aliquots and store at $-85°C$. At the day of analysis, dilute the solution 1/20 in the assay buffer.

11.3.2 Procedure

The assay is adapted to determine either the NO synthase activity directly or by measuring the total levels of nitrites after incubation with nitrate reductase.

11.3.2.1 Total Nitrite Assessment

In a clear microplate, add 50 µL of S15 fraction (15,000 × g supernatant of homogenate), blank, or nitrite standard (5 µM final concentration) with 25 µL of reduced NADPH and 25 µL of nitrate reductase activity. Incubate for 30 min at room temperature, add 100 µL of Griess reagent and read absorbance at 540 nm. Standard solution of nitrates could also be added as a positive control for nitrate reductase activity.

11.3.2.2 NO Synthase Activity

In a clear microplate, mix 50 µL of S15 fraction with 50 µL of arginine substrate, 25 µL of reduced NADPH, and 75 µL of assay buffer. Incubate for 0, 10, 20, and 40 min at 30°C and read the disappearance of NADPH (excitation at 350 nm and emission at 450 nm or the absorbance at 340 nm). A 100 µL portion at $t = 0$ and $t = 20$ or 40 min could be collected and the total nitrite determined by Griess reagent as described previously.

11.3.3 Data Calculation and Expression

For NO synthase activity the concentration is obtained by the following:

$$[A540_{end\ time} - A540_{t=0}] \times 1/time\ (min) \times [5\ \mu M/(A540_{standard} - A540_{blank})]$$
$$\times\ dilution\ factor\ (200/50) = nmol\ NaNO_2/(min \times mL).$$

Please note that the disappearance in the absorbance or fluorescence of reduced NADPH could be used as well.

This value is then normalized against biomass (total protein contents in mg/mL) to obtain nmol nitrites/mg proteins or g tissue weight.

11.4 CYCLOOXYGENASE ACTIVITY

Cyclooxygenase is the enzyme responsible for the oxidation of arachidonic acid to cycloendoperoxide, which is the precursor for prostaglandins, leucotrienes, and thromboxanes. These are pro-inflammatory mediators and they are involved in pain and oxidative stress in vertebrates and invertebrates. Cyclooxygenase activity is inhibited by salicylates and nonsteroidal anti-inflammatory drugs such as aspirin, ibuprofen, naproxen, and mefenamic acid.

$$Arachidonic\ acid + 2O_2 \rightarrow cycloendoperoxide\ (PGG2)$$

The assay is based on the time-dependent elimination of cycloendoperoxide by peroxidase and dichlorofluorescein substrate as described previously.

11.4.1 Reagents

KH$_2$PO$_4$ buffer (50 mM): Dissolve 0.68 g KH$_2$PO$_4$ in 90 mL water, adjust pH to 8 with NaOH 1 M, and complete to 100 mL with distilled water.

Arachidonic acid substrate: Dissolve 6 mg of the acid in 5.4 mL of dimethyl-sulfoxide (6 mM). Store in 500 μL aliquots at −20°C.

Assay buffer: 50 mM Tris-acetate, 0.1% Tween 20, 10 μM hematin, and 0.5 mM EDTA, pH 8. Prepare a 100 mL working buffer. Hematin stock is prepared at 5 mg/mL (8 mM) and diluted to 10 μM in the buffer: 0.125 mL of stock/100 mL.

Hydrogen peroxide standard: Prepare a 10 mM dilution in bidistilled water or 0.02%. Hydrogen peroxide is sold at 10 − 30% concentrate commercially. Prepare daily and keep at 4°C.

Peroxidase reagent: Dissolve 10 mg horseradish peroxidase and 1 mg of dichlorofluorescein in 100 mL of assay buffer. Dilute 1/10 in the KH$_2$PO$_4$ buffer.

Fluorescein standard (10 μM): Dissolve 3 mg in 10 mL of bidistilled water. Dilute 1/100 in assay buffer to obtain a final concentration of 3 μg/mL.

11.4.2 Procedure

Cyclooxygenase is found in cytosol, hence the assay is commonly practiced on the post-mitochondrial fraction ($>$10,000 $\times g$). Mix 10−50 μL of post-mitochondrial fraction or blank (homogenization buffer) and complete to 180 μL with the assay buffer. Start the reaction by adding 20 μL of arachidonic acid substrate and incubate for 0, 10, 20, 30, and 40 min at 20−22°C.

At each incubation time, pipette 20 μL of reaction mixture and mix with 80 μL of peroxidase reagent and incubate for 15 min and measure fluorescence at 485 nm excitation and 520 nm emission. Blanks consist of 20 μL of bidistilled water and a fluorescein standard of 1 μM could be used for calibration (10 μL of fluorescein standard in 90 μL of the assay buffer). The amount of fluorescein will increase over time proportionally with cyclooxygenase activity.

11.4.3 Data Calculation

Cyclooxygenase activity leads to the formation of endoperoxides of arachidonic acid (PGG2), which in turn will oxidize dichlorofluorescein to fluorescein in the presence of peroxidase. Fluorescein fluorescence units could be calculated by the use of fluorescein standard (optional). A standard working solution at 1 μM could be included in the microplate against the blank during the peroxidase reaction. It is calculated as follows:

$$[(Fluorescein\ fluorescence\ units)\ at\ 30\ min]$$
$$-\ [(Fluorescein\ fluorescence\ units\ of\ blank\ or\ t = 0)]$$
$$\times\ 1/30\ min = increased\ fluorescein\ units/min.$$

Fluorescence units could be transformed in nanomoles of fluorescein by multiplying with 1 μM fluorescein/(Fluorescence standard − Fluorescence of blank) giving nmoles of fluorescein/min/mL.

This activity is then normalized against total protein content (mg/mL) or tissue weight (g/mL) to give nmoles fluorescein formed/min/mg proteins or g tissue weight.

11.5 DNA NITROSYLATION

During the process of peroxynitrite production for the destruction of foreign microorganisms by the immune system of the host (phagocytes), the increased presence of NO, which is very reactive to electro-acceptor compounds, could lead to adduct formation of biological macromolecules. Here we describe an assay to determine the levels of NO adducts to DNA molecules in the form of NO_2 adducts. The principle of this assay is based on the enzyme detection of DNA-NO_2 adducts using DNA as a cosubstrate for NADPH-dependent diaphorase. This enzyme catalyzes the reduction of nitrites to amines by

$$R\text{-}NO_2 + NADPH \rightarrow R\text{-}NH_2 + NADP^+$$

In the present case, R is isolated DNA from organisms and previous studies revealed that increased phagocytosis activity in feral clam populations was significantly related to DNA nitrosylation and DNA damage, highlighting the damaging effects of sustained infection to the host [20]. This assay provides a way to determine whether DNA damage is the result of exposure to DNA-damaging compounds or from sustained inflammation (NO production during phagocytosis of microorganisms).

11.5.1 Reagent

Tris-EDTA buffer. Prepare a 10 mM Tris–acetate buffer, pH 8.0, containing 1 mM EDTA.

Phosphate buffer: Prepare a 100 mM solution of KH_2PO_4, pH 7.4.

SDS solution: Prepare a 10% concentration in water.

Phenol-chloroform iso-amyl alcohol: Prepare in the fume hood and in dark glass bottle 25 parts (volume) of phenol, 24 parts of chloroform, and 1 part iso-amyl alcohol. This solution is usually purchased commercially (e.g., Sigma Chemical Company).

Proteinase K: Prepare a 5000–10,000 U/mL solution in phosphate buffered saline (140 mM NaCl, 5 mM KH_2PO_4, pH 7.4).

Ribonuclease A: Prepare a 100 U/mL solution in water.

Diaphorase enzyme: Prepare a 100 U/mL solution in phosphate buffer.

Reduced NADPH: Prepare a 1 mM solution daily in phosphate buffer.

11.5.2 Procedure

DNA from tissues is extracted using a simple universal method developed by Goldenberger et al. (1995) [21]. The homogenate was mixed with SDS at 1% final concentration and proteinase K was added to give 500–1000 U/mL. The mixture was allowed to incubate at 37°C for 30–60 min. After the incubation period, the digested homogenate was mixed with 1 volume of phenol-chloroform iso-amyl alcohol and vortexed. The upper aqueous phase was recuperated and mixed with 1 volume of ice-cold isopropanol. The solution was kept at −80°C for 10 min or 30 min at −20°C and centrifuged at 10,000 × g for 5 min at 4°C to recuperate DNA and RNA. The pellet is resuspended in Tris-EDTA and 10 units of ribonuclease A could be added and incubated at 37°C for 30–60 min (optional) to remove RNA. The material is precipitated again with 2 volumes of isopropanol as described previously. The resulting pellet was dried and resuspended in Tris-EDTA buffer. The purity and amount of the extracted DNA is verified against salmon sperm DNA at 260 (nucleic acid) and 280 nm (protein or phenol). If the ratio is >1.8 then the purity is judged satisfactory. If the ratio is <1.8 then the phenol-chloroform extraction and isopropanol steps should be repeated. The concentration of DNA could be determined at 260 nm using standard solutions (0.5–2 µg/mL) of salmon sperm DNA (Sigma Chemical Company).

DNA nitrosylation is determined as follows. Mix 50 µL of isolated DNA (minimum concentration of 0.1 µg/mL) or blank (Tris-EDTA buffer) with 20 µL of NADPH, 50 µL of diaphorase, and 80 µL of phosphate buffer. The assay is initiated by the addition of reduced NADPH and fluorescence is monitored at each 10 min interval for 40 min at 30°C. Fluorescence readings are taken at 350 nm excitation and 450 nm emission or the absorbance at 340 nm could be taken as well.

11.5.3 Data Analysis

The activity is expressed as follows:

$$[\text{NADPH fluorescence}_{t=0} - \text{NADPH fluorescence}_{\text{end time}}] \times 1/\text{time (min)}$$
$$\times (1/0.05 \text{ mL of added DNA}) = \text{NADPH decrease/min} \times \text{mL}.$$

The activity is then normalized against total DNA in the sample (μg/mL) to give the rate of NADPH decrease/min/μg DNA.

REFERENCES

[1] Brousseau P, Pillet S, Foruin H, Auffret M, Gagné F, Fournier M. Linking immunotoxicity and ecotoxicological effects at higher biological levels. In: Amiard-Triquet C, Amiard JC, Rainbow PS, editors. Ecological biomarkers. New York: CRC Press/Taylor-Francis Group; 2013. p. 131–54.

[2] Auffret M. Bivalves as models for marine immunoecotoxicology. In: Tryphonas H, Fournier M, Blakley B, Smits JEG, Brousseau P, editors. Investigative immunotoxicology. Boca Raton, FL: Taylor and Francis; 2005. p. 29–48.

[3] Canesi L, Scarpato A, Betti M, Ciacci C, Pruzzo C, Gallo G. Bacterial killing by *Mytilus* hemocyte monolayers as a model for investigating the signaling pathways involved in mussel immune defence. Mar Environ Res 2002;54:547–51.

[4] Köllner B, Wasserrab B, Kotterba G, Fischer U. Evaluation of immune functions of rainbow trout (*Oncorhynchus mykiss*)—how can environmental influences be detected?. Toxicol Lett 2012;131:83–95.

[5] Ciacci C, Citterio B, Betti M, Canonico B, Roch P, Canesi L. Functional differential immune responses of *Mytilus galloprovincialis* to bacterial challenge. Comp Biochem Physiol 2009;153B:365–71.

[6] Alvarez-Pellitero P. Fish immunity and parasite infections: from innate immunity to immunoprophylactic prospects. Vet Immunol Immunopathol 1991;126:171–98.

[7] O'Brien M, Mitsuoka T. Quantitative fluorometric assay for rapid enzymatic characterization of Bifidobacterium longum and related bifidobacteria. Microbiol Immunol 2008;35:1041–7.

[8] Douville M, Gagné F, Zhu B, Fortier M, Fournier M. Characterization of commercial microbial products by polymorphic DNA markers and enzymatic activity diversity: occurrence and potential effects on freshwater mussels exposed to municipal effluents. Res J Biotechnol 2010;5:31–9.

[9] Falconer IR, Yeung DS. Cytoskeletal changes in hepatocytes induced by Microcystis toxins and their relation to hyperphosphorylation of cell proteins. Chem Biol Interact 1992;81:181–96.

[10] Al-Tenrineh J, Mihali TK, Pomati F, Neilan BA. Detection of saxitoxin-producing cyanobacteria and *Anabaena circinalis* in environmental water blooms by quantitative PCR. Appl Environ Microbiol 2010;76:7836–42.

[11] Verhaeghe E, Buisson D, Zekri E, Leblanc C, Potin P, Ambroise Y. A colorimetric assay for steady-state analyses of iodo- and bromoperoxidase activities. Anal Biochem 2008;379:60–5.

[12] Lajeunesse A, Segura PA, Gélinas M, Hudon C, Thomas K, Quilliam MA, et al. Detection and confirmation of saxitoxin analogues in freshwater benthic *Lyngbya wollei* algae collected in the St. Lawrence River (Canada) by liquid chromatography-tandem mass spectrometry. J Chromatogr A 2012;1219:93–103.

[13] Kodama H, Baum LG, Paulson JC. Synthesis of linkage-specific sialoside substrates for colorimetric assay of neuraminidases. Carbohydr Res 1991;218:111—9.

[14] Schweiger B, Lange I, Heckler R, Willers H, Schreier E. Rapid detection of influenza A neuraminidase subtypes by cDNA amplification coupled to a simple DNA enzyme immunoassay. Arch Virol 1994;139:439—44.

[15] Odawara F, Abe H, Kohno T, Nagai-Fujii Y, Arai K, Imamura S, et al. A highly sensitive chemiluminescent reverse transcriptase assay for human immunodeficiency virus. J Virol Methods 2002;106:115—24.

[16] Tung JS, Mark GE, Hollis GF. A microplate assay for hyaluronidase and hyaluronidase inhibitors. Anal Biochem 1994;223:149—52.

[17] Ghoneim NH, Abdel-Moein KA, Saeed H. Fish as a possible reservoir for zoonotic *Giardia duodenalis* assemblages. Parasitol Res 2012;110:2193—6.

[18] Almeida A, Pozio E, Cacciò SM. Genotyping of *Giardia duodenalis* cysts by new real-time PCR assays for detection of mixed infections in human samples. Appl Environ Microbiol 2010;76:1895—901.

[19] Lee YC, Yang D. Determination of lysozyme activities in a microplate format. Anal Biochem 2002;310:223—4.

[20] Gagné F, Martin-Diaz ML, Blaise C. Discriminating between Immune- and pollution-mediated DNA damage in two wild *Mya arenaria* clam populations. Biochem Insights 2009;2:51—62.

[21] Goldenberger D, Perschil I, Ritzler M, Altwegg M. A simple "universal" DNA extraction procedure using SDS and proteinase K is compatible with direct PCR amplification. PCR Methods 1995;4:368—70.

CHAPTER 12

Descriptive Statistics and Analysis in Biochemical Ecotoxicology

François Gagné

Chapter Outline

In this chapter, basic methods and approaches of data description and handling are described with biomarkers. It is not intended to be a comprehensive treatise of statistics but to present a general guide of data handling and analysis at the statistical level for the biochemist/biologist in ecotoxicology. When possible, it is recommended to consult with a statistical expert to validate any statistical issues you may have. The procedures are based on the general practice of statistics in the area of biochemical ecotoxicology. More comprehensive and already available statistical manuals and software can be found commercially [1,2]. This chapter presents the most common procedures to describe the data, objectively remove outliers or aberrant values, provide a general outline for statistical analysis based on hypothesis testing and multivariate tools for data mining, and synthesis. In ecotoxicology, the more field-oriented the studies are, the stronger the reliance on statistical models is to identify potential causal relationships. However, the scientist should always be aware of the difference between the statistical versus the biological significance of an observed change. In the ideal world, a significant effect at $\alpha < 0.05$ corresponds to a biological effect in organisms and, conversely, the absence of statistical significance coincides with the absence of biological effects. However, in research we are often confronted with situations where biological effects are measured but not at the statistical significant level or there are no biological effects (i.e., without important consequence), although a statistical difference is found. It is the role of the expert to understand the nuance between statistical and biological significance of effects. The most common example is when an observed biological effect is found but at α values between 0.05 and 0.1. Often called marginally significant effects, it is the investigator's task to consider whether an actual biological change is found in the population. The investigator may want to refute the nonsignificance of the test and decide to repeat and increase replication to judge whether a biological change occurred. This is especially important in field-based studies in contrast with better controlled laboratory tests that strongly rely on statistics to identify relationships or trends.

12.1 DESCRIPTIVE STATISTICS—BACK TO BASICS

Based on Chapter 2, data are generated in replicates at the analysis step (method variation) and at the individual level (biological or interindividual variability). The number of individuals used depends on the expected variation of the responses. For example, in a controlled laboratory exposure experiment where the animal size, age, reproductive state, nutritional status, and sex can be controlled or minimized at least, few organisms are required (between four and eight individuals per treatment group). In the case of field-based studies, where many or some of the above variables are not controlled, the number of animals needed is usually higher (15—30 animals per treatment group, i.e., site or temporal point). The number of replications can be determined by power analysis, which recommends the number of replicates or animals needed to highlight a significant given change (20%, 30%, or 40% changes in respect to controls, for example).

In ecotoxicology, it is generally accepted that data follow a log-normal distribution when the replication is optimal (Figure 12.1). When data follow a log-normal distribution, it is considered parametric. This warrants the use of the mean with a corresponding measure of variation (SD, 95% confidence interval, SE) as a descriptor of the data. An example is provided in Figure 12.2 to highlight the description of parametric data. The distribution of the data appears to follow a trend and shows some degree of symmetry around the mean value.

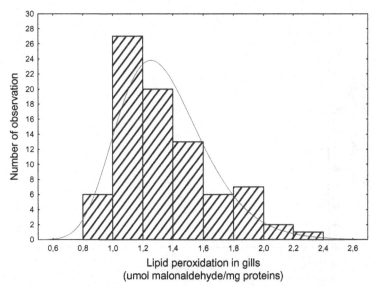

Figure 12.1 Classic example of log-normal distribution of a biomarker. The distribution of LPO in mussel gills as determined by the thiobarbituric acid methodology is shown.

In this example, the central value of the data is expressed as the mean and the variation expressed as the SD, the SE, or the 95% CI (Figure 12.2A). The SE is obtained by dividing the SD with the square root value of the treatment's replicate or organisms ($\sqrt{N}$). The 95% CI could also be used as a measure of dispersion (Figure 12.2B). It is more closely related to the SD (smaller than the SD) than the SE. The CI indicates the reliability of an estimate by which the data distribution would differ significantly from each other. The lower end-point of the 95% CI is as follows:

$$\text{Lower endpoint} = \text{mean} - 1.96[\text{SD}/\sqrt{n}],$$

and the upper endpoint of the 95% CI = mean + 1.96 [SD/$\sqrt{n}$)], where n is the number of replicates. CIs provide insights into the reliability of the data, i.e., whether it falls within the range where 95% of the data are found. Data points that are outside this range could be considered from another (and different) distribution of data.

When the data do not follow a log-normal distribution, i.e., the distributions do not follow a continuous and symmetrical trend, they are considered nonparametric (Figure 12.3A). Nonparametric data do not show a continuous log-normal distribution but rather a more "noisy" distribution of data.

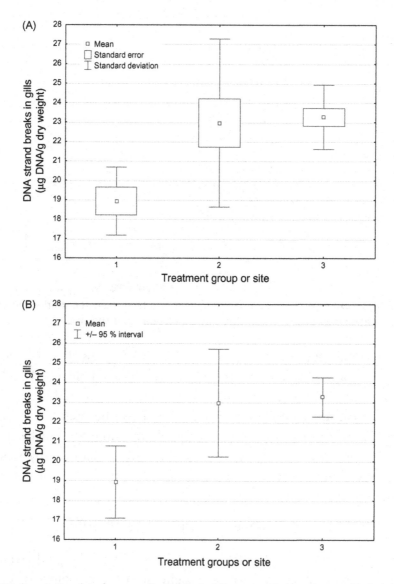

Figure 12.2 Descriptive data for normally distributed (parametric) data. Parametric descriptor of the biomarker is the mean with a measure of variation such as SD/error (A) and 95% CI (B). CIs that do not overlap are usually statistically different at the $p < 0.05$ level.

In Figure 12.3A, the data clearly display a distribution lacking symmetry and differ from the log-normal or normal distributions. In many instances these types of distributions hide uncontrolled variables such as gender (males or females or undifferentiated sex) and size-related effects: small versus large

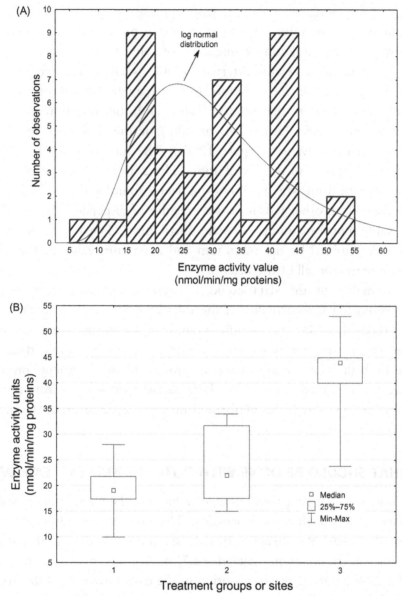

Figure 12.3 Descriptive statistics for nonparametric data. An example of nonparametric distribution is shown (A). Nonparametric data are usually described by the median value with either the 25th and 75th percentile or the minimum and maximum values (B).

organisms, age (young vs. old), or genetic make-up (species or subspecies). This could also be the result of data expressed as proportions or when ranked in some way. In this situation, it might be of value to reconsider the experiment and attempt to tease out external variables such as size, weight, age, or gender when possible. In the situation of nonparametric distributions, the data are usually expressed as the median value and the variation of the data expressed as percentiles (25th, 75th, or 50th) or the difference between the minimal and maximal values (Figure 12.3B). The median value is defined by the value at the middle of the distribution. The 25th and 75th percentiles represent the interval from which 75% of the data is found and is considered analogous to the SD; in other words, it is a measure of dispersion around a central (middle) value, but for a nonparametric distribution. The whole dispersion of the data is determined by the minimum and maximum value of the distribution, which represents all of the data.

The normality of the data could be determined statistically using the Shapiro—Wilks test for normality. If the calculated W value is deemed significant ($\alpha < 0.05$) then the distribution is considered nonparametric or not normal. One common error is to test the normality of the whole data, which comprise both the control and treatment groups. If the treatments produce an effect then it is expected to obtain a whole dataset that deviates from normality. The normality tests should be performed on each separate group (control and treatment).

12.2 WHAT SHOULD BE DONE WITH OUTLIERS AND EXTREME DATA?

In some instances, the data could contain values that lie outside the distribution of the data while still showing normality. The presence of outlier or extreme values could "smear" the variation of the data and mask any significant results. For example, in the following dataset—2.7, 3, 3.3, 4.5, 5.1, 5.5, and 10—the value 10 is clearly outside the distribution of the data. However, if the "outside" data were 6.3 or 6.8, it becomes less apparent to consider it as part of the data distribution or as an outlier. An objective way to determine this is to use the following rules depending on the type of distribution (i.e., normal or not normal).

We define outlier values if the datum is outside this range:

Datum point > upper value of SE + 1.5 × (upper − lower value of SE) or

$$\text{Datum point} < \text{lower value of SE} - 1.5 \times (\text{upper} - \text{lower value of SE})$$

For nonparametric data, the upper and lower values of the SE are replaced by 75th and 25th percentiles, respectively. Extreme values are defined when data fall outside with a factor of 2 instead of 1.5. This could be represented visually to highlight these definitions (Figure 12.4).

Extreme data could also be the result of a "bad day" for a research student or post-doctoral researcher running the assays or even worse, when the senior researcher is performing the assay instead... To "clean" the dataset objectively, the extreme values (*) could be first removed from the dataset by following the previous rules. However, the threshold should be recalculated each time a datum point is removed from the set. For outlier values, it is debatable that the data should be removed from the dataset especially when <6 replicates are used. The important thing is to be consistent (objective) in doing so across all treatment groups. From my personal experience, when the outliers are not

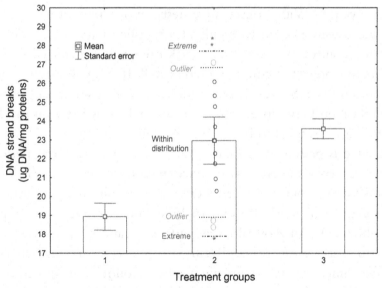

Figure 12.4 Identification of outlier and extreme values in data. Data points within the distribution (blue open circle) are within the upper or lower value of the SE ± 1.5 × (upper − lower value of SE). Data points outside this range are either outliers (orange open circle) or extreme values (*).

symmetric in respect to mean, i.e., a high and low data point with the mean or median between them, the point(s) are removed from the dataset. For example, if the outliers are either on the upper or lower threshold, they could be removed if the number of replicates is ≥ 6. Conversely, if the outlier data are found at both the upper and lower limit, then they are maintained in the dataset.

12.3 UNIVARIATE STATISTICS

12.3.1 Hypothesis Testing

Univariate statistics consist of determining whether a treatment group produces a significant difference against the controls or between different treatments altogether. This represents classical hypothesis testing when only one effect (treatment group) is considered as the main driver of effects. When two groups are used, multicomparison tests are used such as the t test. When more than two groups are used (three groups or more), analysis of variance (ANOVA) is used first, followed by a multicomparison test such as the least-square (LSD) test, Dunnett's or Bonferroni's t test, and Scheffé's F test. These tests differ in "conservatism," i.e., some are least severe than others. For example, the LSD test is considered the least conservative test while the Scheffé's or the Tukey's test is more conservative. With conservative tests, more replication and stronger responses are usually needed to highlight a significant difference. When comparing with a control group, the Dunnett's t test is the preferred test by some software but any other test could be used as well. If an uneven replication exists between groups the HSD test is recommended, which was designed to handle unequal replication between groups. This test, which is between the LSD and Scheffé's test, is considered to be moderately conservative.

An example is presented in Figure 12.5. An experiment was performed in which mussels were exposed to two concentrations (1 and 10 µg/L) of xenobiotic A for 10 days. Control mussels were exposed only to aquarium water and with the solvent vehicle (0.01% dimethylsulfoxide) if required to properly disperse xenobiotic A. An ANOVA was performed and gave a significant result ($p < 0.05$) so that we can proceed with multiple comparisons between treatment groups. Since we are comparing against a control (unexposed to xenobiotic A), the Dunnett's t test was used to highlight significant changes between the control (0) and treatment 1 or 10. Treatment 10 µg/L gave a significant result while treatment 1 µg/L did not produce significant changes.

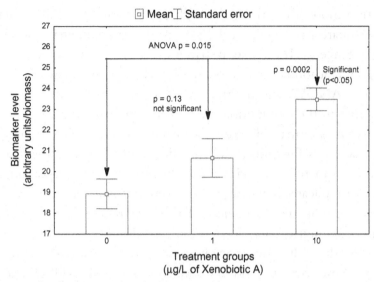

Figure 12.5 Example of ANOVA and multicomparison tests.

However, before using these tests, the following two assumptions must be met. First, the data within treatments should be normally distributed, which can be determined by the Shapiro—Wilks test for normality. Again, a common error at this step is to apply the normality test across all three treatment groups, which could show an asymmetric or abnormal distribution because treatment 10 will introduce a different distribution in the dataset. In other words, when a treatment introduces a significant change, it will bring a different distribution in the dataset, which gives an abnormal result. The normality should be checked within the controls and within each treatment group. The second assumption is that the variances of each treatment group should be "homogenous," i.e., they are similar in proportions in respect to the mean value of the data distribution in each treatment group. This is conveniently verified by either the Levene's test or Brown—Forsythe's test for homogeneity of variance. The latter is considered more robust than the former because it can handle stronger deviations of normality or unequal replication between groups.

In situations when the data are not normal or they present a heterogeneous variance between treatment groups, the analyst has two options. The first option is transforming the data to limit the spread of the distribution, which gives a distribution closer to normality. This is achieved by logarithmic ($\log_{10}$)

transformation most of the time, although other transformations could be used such as the square root, reciprocal (1/data), or trigonometric transformations (sine, cosine, tangent). If the data distribution is still nonparametric (parametric ANOVA could accommodate moderate deviations from normality) then a nonparametric ANOVA and multiple comparison tests are used. In this procedure, the values in the distribution are ranked before performing the ANOVA. Usually software automatically runs nonparametric ANOVA (Kruskal–Wallis ANOVA), but if this function is not included in the software the analyst could first rank-transform the data within each treatment (it is important here that the number of replicates per treatment is constant) and then run the ANOVA. In other words, rank-transformation of the data represents the last resort for handling nonparametric data distributions. Multiple comparisons are usually performed by the following nonparametric tests: Mann–Whitney U test, Kolmogorov–Smirnov test, and the ranked based Tukey-HSD test. As seen earlier, these tests vary in conservatism where the Kolmogorov–Smirnov and the Tukey-HSD tests are the least and the most conservative tests, respectively. The Mann–Whitney is mathematically similar to the t tests (i.e., when applied to normally distributed data, it is equivalent to the Student t test) and is considered moderately conservative.

12.3.2 Correlations

Correlation analysis is performed to identify the strength of relationships between a pair of variables. The correlation coefficient r varies between -1 and $+1$ where a perfect correlation is ± 1 and 0 is the absence of correlations. Values of r between 0 and 1 reflect a partial correlation, which can be significant or not. For example, $r = 0.80$ indicates that variable 1 is related to variable 2 at 80%. In some cases, the squared value of r is applied to always have a positive value and is defined by R or r^2. Only correlations that are significant at $p < 0.05$ or 0.01 should be considered. When the data are parametric and normally distributed, Pearson-moment correlation is used. Linear regression is similar to linear correlations, but it is assumed that one variable (dependent) depends on another independent variable. The slope in the linear regression equation is given by linear equation $Y = AX + B$ and is produced using the least-square method where a line is placed in the data plot that gives the least difference (minimize the error between the fitted line and the data points) between the line point and the actual data point. The squared difference is measured to remove the negative sign. Slope A is obtained from the ratio of $Y_1 - Y_0/X_1 - X_0$ in the line. The slope and correlation coefficient are similar, but the correlation coefficient is scaled to fit between 0 and 1 while the slope depends on the units of y and x. An example is shown in Figure 13.6. In this figure, a linear relationship is extracted using the x and y data points. The fitted line is placed at the least sum of errors between the data points and the fitted points from the linear line. The line produced a correlation coefficient $r = 0.88$ and significant at $p = 0.01$ level.

For example, increased exposure to cadmium (Cd) involves the production of reactive oxygen intermediates which, uncontrolled, could lead to oxidative stress such as lipid peroxidation (LPO). Cd is detoxified upon binding to cysteine-rich proteins called metallothioneins (MTs). MTs are inducible by exposure to Cd and by oxidative stress. One way to determine whether increased MT levels are associated more to LPO (i.e., oxidative stress) than to Cd is to examine the correlation between these two biomarkers. If a significant correlation is found then it supports statistically the argument that increased MT levels are associated with increases in oxidative stress (LPO).

In some cases, the residual value of linear regression between two variables could provide insights on interactions between biomarker responses. In Figure 12.6, the difference of the Y biomarker between the fitted line and the

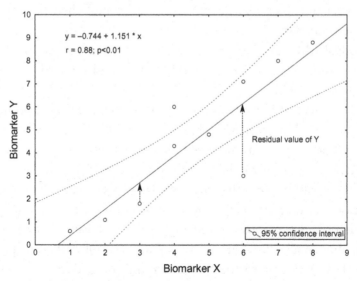

Figure 12.6 Example of a linear correlation between biomarker X and Y.

data point represents the error or the residual value of biomarker Y. The residual value of biomarker Y could be interpreted as the portion of the biomarker that is not explained by the linear regression with biomarker X, i.e., independent of biomarker X. This approach forms the basis of analysis of covariance (ANCOVA) explained in the following section. This approach could be used as a way to normalize a biomarker response with the amount of added sample (biomass; see Chapter 2). If the changes in biomarker responses are solely related to the added biomass, then a straight and highly correlated line (with low error) between the biomarker level (y-axis) and the biomass value would be obtained. However, if some biomarker responses are affected by other factors than biomass, such as the treatment groups, then deviation from linearity would be observed. In other words, the error between the linear line and the data point represents the independent part of the biomarker against the biomass and is related to the effects of the treatment group (at the hypothesis testing level). For example, the increased exposure to Cd in fish liver cells (hepatocytes) is related to the expression of MTs and oxidative stress in cells. Given that Cd-bound MT is less toxic than free Cd, one would want to determine whether oxidative stress is more related to MT expression or to exposure to Cd. If oxidative stress is significantly correlated with MT levels at $r = 0.55$, then oxidative stress is related not only by exposure to Cd but to the expression of MT as well. If we plot the residual value of MT from the MT versus oxidative stress correlation plot against the treatment groups then the following pattern could be observed:

1. The residual MT values do not change across the increasing Cd exposure concentration: we could say that MT is increased by oxidative stress rather than exposure to Cd.
2. The residual MT values change with the increasing Cd exposure concentration: we could say that MT changes are related to Cd exposure concentration and not only from increased oxidative stress.

The residual extraction method could be used to tease out (linear) interactions between variables when they are correlated with each other.

In the case of nonparametric data distributions, the data could be transformed as described previously using $\log_{10}$ and root values of the data. If the data remain abnormally distributed, the values are ranked before performing a correlation. This is similar to the Spearman rank procedure for nonparametric data. In some cases, when the number of data points is low ($n = 4-8$),

Spearman rank correlations are more suited to determine correlations between pairs of variables.

12.3.3 ANCOVA

In some situations, the treatment groups could have mitigating effects on the expression of biomarker Y. The treatment groups could have a direct effect on another biomarker (X), which in turn is highly associated to biomarker Y. It is therefore difficult to determine what bearing the treatment groups have on biomarker Y. ANCOVA represents a statistical means to tease out these interactions and in some instances, decrease the between–group error (or variability). It is based on the residual extraction method explained previously in which biomarker Y corrected against biomarker X is compared with the treatment groups. ANCOVA combines ANOVA with linear regression where the biomarker responses are examined in terms of the treatment groups and corrected against covariables, which are linearly correlated with the biomarker responses. It is similar to running an ANOVA on the residual value of the biomarker obtained from a linear relationship of the biomarker with other variables (another biomarker, for example). ANCOVA has the potential to increase the power of the experiment because residuals have lower variation than the biomarker response (i.e., it decreases the interindividual/group variation). However, ANCOVA procedure depends on many assumptions:

- The residuals should be normally distributed, thus, not applicable to nonparametric distributions.
- The variance of the error should be equal across treatment groups.
- The slopes between each regression line across the treatment groups should be similar (i.e., parallel).
- The error terms should be uncorrelated with each other.

Returning to the previous example where fish hepatocytes are exposed to increasing Cd concentrations (treatment groups), the levels of MT are increased with the Cd exposure concentrations. The levels of LPO were also determined and were significantly correlated with MT levels in cells. One could ask whether LPO or Cd exposure is the main driver of MT expression in cells. An ANCOVA is then used to determine whether the covariate would be LPO, the test variable MT, and the treatment groups as the factor. If the p value of the covariable is higher than the p value of the main factor (Cd exposure concentration) then the contribution of the covariable is more important than the treatment group. If the residual MT levels are highly significant with low or no significance with the covariable, then the contribution of LPO on the expression of MT is considered negligible.

12.4 MULTIVARIATE STATISTICS

In ecotoxicology, a series of biomarkers are often measured along with bioaccumulation measurements of various contaminants in tissues. This leads to a complex dataset where the relationship between each endpoint becomes unclear and also if some or all of the above measurements are characteristic to the treatment group or site location (spatial survey) or point of time (temporal survey). This situation is often observed with field-based studies or studies dealing with high-throughput assays such as the "-omics" approaches.

12.4.1 Factorial Analysis

Factorial analysis is performed when one wants to find which biomarkers (or a subgroup) best explain the variance of all biomarkers in a complex dataset. For example, if a research group measured the expression of 50 genes in mussels collected at 10 sites upstream and downstream an urban area, one could ask whether a subgroup of gene expression changes could explain the overall responses and whether some genes are more important than others. In other words, is there a subgroup of endpoints that is correlated with n-dimensional vector composed with the 50 genes of interest? This analysis could be considered as a data reduction method in finding which biomarkers best explain the whole dataset. In some cases, no subgroups of biomarkers could be found and the whole dataset should be kept to examine the changes across 10 sites. In many instances, a subset of biomarkers could be proposed that would reduce cost and time to study the effects across the 10 sites in an urban area. Factorial analysis is also a methodology to examine the inter-relationships between biomarkers, which could provide interesting insights to better understand the interaction between the toxicity of pollution at the biochemical/physiological level.

An example of factorial analysis is given in Figure 12.7. This example is based on the production of 11 biomarkers in organisms exposed to various treatments or collected at different sites. The biomarkers could be contaminant tissue burdens and any biomarkers to determine organism's health and toxicity. Factorial analysis using the principal component procedure is the most

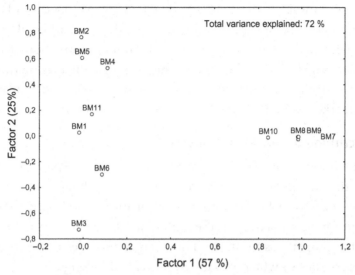

Figure 12.7 Example of factorial analysis results using the principal component procedure.

commonly used method, but other methods can be used, such as centroid, principal axis, and multiple R^2. These methods for extraction of main factors are available in many types of statistical software. The method determines principal factors, which are eigenvalues obtained from n endpoints (biomarkers). The first two factors, which explain most of variance, are plotted on an X and Y plot. The percentage value on the X- and Y-axis are the proportion of the variance explained where the total is indicated in the plot (72%). In practice, if the variance is explained at more than 60% then the analysis could be considered acceptable for identifying the most relevant biomarkers. If the total explained variance is below 60% then the biomarker used cannot explain all the variations in the data. In the figure, factorial weights (or the correlation of the biomarker with factors 1 and 2) of the biomarkers (BM1−11) are plotted in a 2D plot. Biomarkers with factorial weights >70% are considered the most important biomarkers or principal components because they are more strongly correlated with the eigenvalue of factors 1 and 2. In the example, these are BM2, 3, and 7−10. This indicates that 72% of the data could be explained by 6 of the initial 11 biomarkers. In these plots, biomarkers grouped together (e.g., BM 7−10) are generally positively correlated with each other while the ones that are diametrically opposed (e.g., BM2 and 3) are often negatively correlated. This provides information on the inter-relationships between biomarkers; for example, if we examined the biomarker of toxicity in mussels collected at 10 sites upstream and downstream a polluted area—the biomarkers that are related to polyaromatic hydrocarbons in tissues and biotransformation activity show high factorial weights—then the analysis suggests that these compounds are the main culprit in the contamination source. Of course, this is defendable if other types of biomarkers were examined.

12.4.2 Discriminant Function Analysis

Discriminant function analysis is similar to multivariate ANOVA but indicates how well the treatment groups or study sites differ with each other. Discriminant analysis seeks out a linear combination of biomarker data for each treatment group that maximizes the difference between treatment groups or study sites for proper classification. The analysis provides a classification function that determines to which groups an individual belongs:

$$s_i = c_i + w_{i1}{}^*x_1 + w_{i2}{}^*x_2 + \cdots + w_{im}{}^*x_m$$

where i represents the respective groups or sites and numbers 1,2, ..., m represent the m biomarkers (variables). c_i is the constant for group i and w_{ij} is the ponderation factor of variable (biomarkers) j for group i. S_i is the classification value. Different sites will give classification efficiency greater than 70% while similar sites would give undistinguishable results (<70% classification). The analysis provides a classification matrix that shows how well each individual or treatment group replicate is associated to a given site or treatment group. The results for discriminant function analysis are usually expressed by the Wilks' lambda statistic and displayed by a 2D plot of the two best discriminant functions in respect to site classification (Figure 12.8). The analysis also provides factorial analysis to identify which biomarkers are strongly correlated with the discriminant functions 1 and 2. In the example, 11 biomarkers were examined in six sites (groups 1—6) in local mussel populations. In this analysis, each data point represents the root function of the combined 11 biomarkers for each individual. The data points are also displayed for each treatment group or site of collection. In the legend, the classification results are presented in percentage correctness. For example, group or site 2 was correctly classified at 100% but sites 3 and 5 were misclassified as site 2. This indicates that sites 2, 3, and 5 could be considered a similar "composite site," or these sites could be

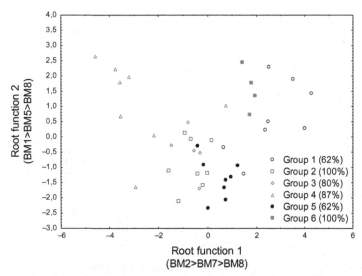

Figure 12.8 Example of discriminant function analysis for site classification. Eleven biomarkers (BM) were determined in six groups (sites or treatments) and analyzed by discriminant function analysis. Each data point corresponds to each replicate individual in a group. The percentage values of groups 1–6 represent the classification correctness. Groups that overlap with each other are considered similar, while distinct groups are considered different. BMs that were the most strongly correlated with the root functions 1 and 2 are mentioned on the x- and y-axis.

considered similar in their ecotoxicological properties. Group 6 was correctly classified at 100% although close to site/group 1. The biomarkers that had the highest correlation with the root function are listed in parentheses on the x- and y-axes. This type of analysis is often used in ecotoxicology because it provides a great deal of information for data mining and analysis. Indeed, it provides information on how sites differ with each other (if sites are similar in nature or not) and which biomarkers contribute to site classification/discrimination.

In discriminant function analysis, the factorial scores could be calculated for each observation, which is a linear combination of the standardized biomarker measured for each individual (Figure 12.9). It provides an integrated metric that reflects the changes in the 11 biomarkers tested. The larger the factorial weights (absolute value) the greater the contribution of the biomarker to the root functions 1 and 2.

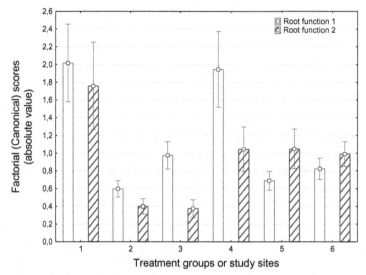

Figure 12.9 Factorial scores of the biomarker data. These data are expressed as the mean of the absolute scores with the SE.

12.5 ARTIFICIAL NEURAL NETWORKS

Artificial neural networks (ANNs) are mathematical algorithms that simulate the learning processes in nerve tissues composed of neural networks (Figure 12.10). The power of an ANN-based model resides in its capacity to "learn" about relationships within the data that go beyond linear relationships (nonlinear modeling). As with discriminant function analysis and factorial analysis, ANNs are powerful classifiers and they are used to seek out trends in complex and noisy datasets. ANNs based on back propagation learning are by far the most used ANNs for predicting changes in the field of ecotoxicology. ANNs have been used in many applications in field and laboratory studies such as cancer diagnostics, genetics, oceanography, psychology, environmental sciences, and ecotoxicology. For example, Budka et al. (2010) [3] used biomarkers in mussels to classify coastal water quality. In this study, a suite of biomarkers of oxidative stress—such as LPO, catalase, and total oxygen radical scavenging capacity, among others—in blue mussels was successfully used to classify the water quality near the coastline.

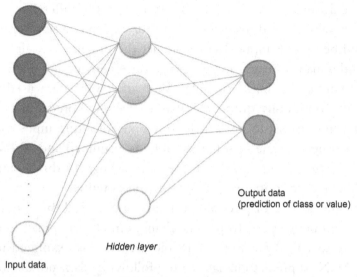

Input data

Hidden layer

Output data
(prediction of class or value)

Figure 12.10 A back propagation ANN. Back propagation neural networks are trained with the input data in respect to the output data. The interconnections pass through one or more hidden layer(s) to generate the output data with the least error possible. Learning is achieved by forward and backward adjustment to minimize the error between the predicted output data and the actual output data during learning. Learning is usually achieved between 10 and 200 cycles, depending on the complexity of the data responses.

A unique feature of ANNs is that they are trained and developed with the owner's dataset. The strength of the network connections is built by a learning process in which input data are trained in the presence of output data (to be predicted). By the process of going back and forth (back propagation learning scheme), the strength of the networks is adjusted to find an arrangement providing the least error in the output data. In doing so, generalizations are made that could reveal unsuspected or unforeseen trends in the various biomarkers in large datasets. ANN models are included in many statistical software packages (as optional attachments) and have three major purposes: classification, prediction, and forecasting from a large and complex input dataset [4,5]. In classification, the input variables are examined in terms of different categories as the output data. For example, a suite of eight biomarkers could be used to identify in which class a given organism is associated with three states: (1) normal/healthy condition, (2) stressed by pollution, and (3) harmful or pathological condition. Prediction is similar to classification with the difference that a continuous output value is used instead of class. For example, a suite of biomarkers of stress could be used as input data to predict the organism's weight to length ratio (size) and gonado-somatic index (gonad development and growth). In this case, a set of continuous input variables (biomarker value) are used to predict another set of continuous output biomarker values. Forecasting consists of using a given input dataset to predict the outcomes of future input datasets. For example, the change in time of input variables is examined and ANN could be developed to predict changes in the future (forecasting). In this situation, other variables could be used to increase the prediction potential.

As an example, a suite of physiological and biochemical markers was examined in feral clam populations to predict changes in clam population characteristics [6]. More specifically, a suite of 18 biomarkers was examined using back propagation ANN to predict changes to the following clam population characteristics: live clam number, total clam number, empty shells (dead clams, predation), and sex ratio. In an ANN software package, a data reduction function is included that is similar to data reduction using factorial analysis. This is sometimes called a genetic selection algorithm or sensitivity analysis of input data, which computes the error induced in the prediction of the output data when a given input biomarker is left out during the training phase. In doing so, the software identified 6 of 18 biomarkers that could effectively predict output data on clam population characteristics: gonado-somatic index, gonad maturation, mitochondrial activity at 4°C, LPO, DNA strand breaks (digestive gland), and phagocytosis. This does not mean that these endpoints are actually those responsible for decreased clam number and death incidences but they are those that best predict the changes. Another ANN was produced consisting of the six selected input biomarkers and produced a network with less than 3% error prediction for the training data (seen data) and 20% error with unseen data. This suggests that the earlier biomarker endpoints could serve to predict impacts at a higher level of biological organization (population clam metrics). Decreased gonad maturation and growth, altered immunity, oxidative stress, and energy expenses are indeed physiological endpoints that could impact clam number, clam death or susceptibility to predation, and altered sex ratio, but this should be examined more closely to confirm this at the study sites for confirmation or validation. In the context of monitoring surveys, these physiological targets could be considered when measuring potential ecotoxicological effects in feral clam populations.

REFERENCES

[1] Gad SC. Statistics for toxicologists. In: Hayes AW, editor. Principles and methods of toxicology. fifth ed. New York: CRC Press; 2007.
[2] Bailer JA, Piegorsch W. Statistics for environmental biology and toxicology. London: CRC Interdisciplinary Statistics; 1997.
[3] Budka M, Gabrys B, Ravagnan E. Robust predictive modelling of water pollution using biomarker data. Water Res 2010;44:3294–308.
[4] Khan J, Wei JS, Ringner M, Saal LH, Ladanyi M, Westermann F, et al. Classification and diagnostic prediction of cancers using gene expression profiling and artificial neural networks. Nat Med 2001;7:673–9.
[5] Sahoo G, Ray C, Mehnert E, Keefer D. Application of artificial neural networks to assess pesticide contamination in shallow groundwater. Sci Total Environ 2006;367:234–51.
[6] Gagné F, Blaise C, Pellerin J, Fournier M, Durand MJ, Talbot A. Relationships between intertidal clam population and health status of the soft-shell clam *Mya arenaria* in the St. Lawrence Estuary and Saguenay Fjord (Québec, Canada). Environ Int 2008;34:30–42.

CHAPTER 13

Biomarker Expression and Integration

François Gagné

Chapter Outline

The purpose of this chapter is to present different means of expressing toxicity endpoints and an introduction to the integrated biomarkers index. Toxicity data are often expressed in terms of either threshold concentrations or effects concentration derived from linear concentration or dose—response curves for the latter. In toxicology, most dose—response relationships are linear or sigmoid (Figure 13.1A). The effects concentrations such as the EC_{20} (the concentration that produce a 20% change) are derived by linear regression. For biomarkers, dose—response or concentration—response curves often present an inverse U-shaped response curve (Figure 13.1B) where the endpoints are expressed at low concentrations and return to control values as the exposure concentration increases. In some situations the dose—response curves with biomarkers often show a hormetic response or inversed U-shaped dose—response curve. There are many reasons behind this behavior, but most often the inverse U-shaped curve is the result of either saturation of the biochemical pathways involved in the expression of the given biomarker or decreased expression because of cytotoxicity where the cells gradually lose their ability to express the biomarker as the concentration increases.

In some cases, the inversed U-shape curve goes down to even below the controls where it is considered a hormetic concentration/dose—response curve (Figure 13.1C). Hormesis was found in many instances where an initial

Biochemical Ecotoxicology
DOI: http://dx.doi.org/10.1016/B978-0-12-411604-7.00013-1

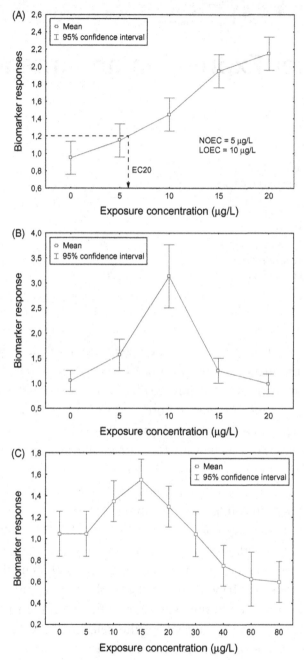

Figure 13.1 Typical concentration—exposure concentrations with biomarkers. Linear (A), inversed U-shaped (B), and hormesis type (C) of exposure—response curves are shown.

increase in the response was observed at low concentrations followed by decreases at higher concentrations [1]. Hormetic responses are gaining more and more recognition and take into account biological plasticity or variation (if not duality) in the organism's capacity to respond to external physical and chemical agents [1]. It was found that hormetic dose—response relationships were independent of the biological model, endpoint measured (effects), and the nature of the physical/chemical agent [2].

These observations lead to the realization that exposure—response relationships are complex and can often deviate from linearity, thus complicating the description of toxicity data based on simple linear relationships. Hence, the two descriptors of toxicity described previously (i.e., thresholds and EC_x values), representing the two classical ways to report toxicity data and risk assessment, could be questioned. The use of a descriptor of toxicity that is independent of the model (i.e., linear or sigmoid equations or sinusoidal hormetic responses) and integrates both the threshold concentration and the intensity of the responses would be of value. An approach based on the method of variation is also proposed to handle complex dose—response relationships.

13.1 TOXICITY ENDPOINTS

13.1.1 Threshold Concentrations

Concentration thresholds provide information on the concentration of a given xenobiotic or mixture in terms of the lowest observed effects concentration (LOEC) and no observable effects concentration (NOEC). It does not provide information on the intensity of responses, only responses that differ statistically from the controls. The determination of LOEC is based on statistical procedures to find by which a given exposure concentration produces a significant effect that calls for analysis of variance (ANOVA) and *post hoc* tests, such as the Dunnett's t test, when comparing increasing concentrations of a test compound against the unexposed or control group. The NOEC concentration is the concentration just before the LOEC that does not produce a significant change. The LOEC and NOEC will give the concentration of a test substance that will produce a significant effect in the organism. In some studies, a threshold concentration is calculated from the geometric mean between the NOEC and LOEC: $(NOEC \times LOEC)^{1/2}$. Although this approach is widely used in environmental toxicology studies and risk evaluation of xenobiotics, threshold concentrations are limited by the a priori selection of exposure concentration. As already mentioned, this method lacks information on the intensity of the observed effect. For example, the toxicity of a given xenobiotic will differ if the xenobiotic produced, e.g., a fivefold increase in the response compared to 1.5-fold increase of the substance at the same effects concentration.

As an example, the LOEC and NOEC concentrations are shown in Figure 13.1A. ANOVA was significant at $p < 0.01$ and Dunnett's t test between the control and 10 µg/L was significant at $p < 0.05$, thus giving an LOEC value of 10 µg/L. There was no statistical difference between the controls and the 5 µg/L treatment group, thus giving the NOEC value. Some apply a threshold concentration that is the geometric mean of the LOEC and NOEC values: $(LOEC \times NOEC)^{1/2} = 7.1$ µg/L. Hence, the threshold concentration is 7.1 µg/L for this hypothetical substance. This is of interest but it provides no information about the "potency" of substances. For example, the substance increases the biomarker response (y-axis) ~ 1.6-fold in respect to controls (Figure 13.1A). Another substance could have a similar threshold concentration but increases the response up to fourfold in respect to controls. Clearly this substance would be more toxic than the former one. A way of accounting for the potency of substances is to determine the effects concentrations based on the exposure—concentration (linear) relationships.

13.1.2 Effects Concentration by Linear Dose—Response Relationships

The use of toxicity data based on the concentration—response relationships has the advantage of including both the threshold effects and the intensity of the observed responses. These are usually expressed in terms of the effective concentration (EC_x), which produces a 20% or 50% change in the response. Historically, the EC_{50} endpoint was used for acute toxicity data (mortality) and was expressed as the lethal concentration that kills 50% of the test organisms (LC_{50}). With sublethal effects biomarkers, the EC_{20-30} is usually used. In Figure 13.1A, the EC_{20} was calculated using the linear relationship. The correlation coefficient was $r = 0.96$ ($p < 0.001$) with the linear equation:

Exposure concentration $= -11.9 + 14.4$ (Biomarker reponse).

A 20% increase of the controls corresponds to $0.95 + (0.95 \times 0.2) = 1.14$. Based on the linear equation, 1.14 gives an exposure concentration of 4.5 µg/L with $\pm 95\%$ confidence interval (CI) of 3.3—5.8 µg/L. This value is close to the LOEC value calculated earlier. Since the value was derived from the slope of the exposure—concentration relationships, this value takes into account the strength (potency) of the chemical to produce a 20% change. Moreover, the calculation of EC_x or LC_x values is dependent on the relation model, which is usually assumed linear or sigmoid (Hill equation). Nonlinear sigmoidal relationships could be modeled by the Hill equation, which is based on cooperativity and saturation interactions. However, current evidence reveals that some compounds such as endocrine disrupters often display non-monotonic responses [3]. Indeed, sublethal effects often show different concentration—response relationships such as U-shaped and quantal responses, which complicates the extrapolation of the descriptor of toxicity. With U-shaped or hormetic concentration—response curves, toxicity data calculations based on these relationships are more difficult to use. In some cases, the linear portion could be considered but the number of data points are greatly reduced in this area. In this case, a different approach to report toxicity data is proposed that is based on the principle of (toxicological) variation.

13.1.3 Method of Variation

The method of variation consists of determining the concentration by which a xenobiotic would induce a variation in the measured biological/biochemical response in organisms exposed to increasing concentrations of a given xenobiotic. This approach takes into account the variation of the measured biological (toxicological) endpoint such as growth, survival, the expression of biomarkers in tissues, etc., instead of the biomarker response itself. The method takes into account both the intensity of the responses and the exposure concentration but makes no assumption on the exposure—concentration relationships.

To explain this, the variation of the effects is associated with a variation of the xenobiotic concentration used (Figure 13.2A). When a compound is devoid of toxic properties, the observed variation is related to the "normal" homeostatic response of the measured effects. With a toxic compound, the same change in the exposure concentration will lead to important changes in the biological response leading to increased variance (Figure 13.2B).

The extent of variation in the effects to the change in the exposure concentration is given by a proportionality factor k, which is a characteristic of the xenobiotic and the measured biological effect. The value of k is defined by

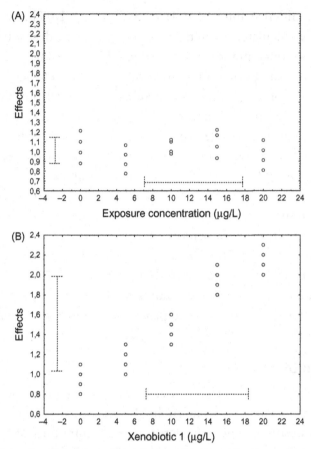

Figure 13.2 Principle of variation as a descriptor metric of toxicity. Exposure to concentrations of nontoxic xenobiotic (A) is associated to the normal variation of the biological response compared to the same variation in the concentration of a toxic xenobiotic (B), which leads to an increase variation in the biological response or effect. The dotted lines represent the SD for the biological effects (y-axis) and the xenobiotic concentration (x-axis). The increase in variation of effects is proportional to the increase in concentration (variation) for a given xenobiotic and is denoted k.

the variation of the biological effects and the variation in the exposure concentration:

$$k = [E_{sd} \times C_{mean}]/ \qquad (13.1)$$

$$[E_{mean} \times C_{sd}],$$

where C_{sd} corresponds to the SD of all the concentrations within replicates and within treatments, and C_{mean} is the mean concentration of all exposure concentrations (including the replicates). E_{mean} is the mean value of the observed effects across all exposure groups, and E_{sd} is the SD (variation) of the observed effects across all exposure groups. Hence, the proportionality k factor is a function in the coefficient of variation of the effects and of the applied concentrations. The k factor is characteristic of the properties of the tested chemical and the physiological nature of the effect. This proportionality constant k will depend on the gradient of exposure regime used, such as linear, semilogarithmic, and logarithmic exposure concentration. k values tend to be higher with linear addition exposure gradient compared to logarithmic distribution: $k_{linear} = 0.85$ compared with $k_{log} = 0.28$. Semilogarithmic exposure concentrations have k values closer to the logarithmic k values: $k_{semilog} = 0.32$. This approach will permit us to derive a descriptor toxicity that takes into account the threshold concentration and the intensity of the response with no assumptions on the type of concentration−response relationship(s). The concentration by which a biological variation is observed (C_{var}) is defined as follows:

$$C_{var} = (C_{mean})^{1/2}/K \qquad (13.2)$$

If $C_{var} \geq C_{mean}$ then the substance has no effects,
If $C_{var} < C_{mean}$ then the substance induces a biological variation.

C_{var} is related to the mean value of the applied concentrations and the proportionality factor k. This new descriptor of toxicity could be simply interpreted as the concentration of a xenobiotic that is associated to a variation in the biomarker response. To illustrate further this concept of biomarker variation as a descriptor of toxicity, several representative examples will be shown. In Figure 13.3, classic linear dose−response (A) and no effect dose−response (B) relationships are shown. In Figure 13.3A, the tested compound is clearly toxic compared to the one in Figure 13.3B. In Figure 13.3A, analysis of the data revealed that the LOEC and NOEC values were, respectively, 10 and 5 µg/L. The EC_{20} (the concentration that produces 20% changes) was calculated at 6 µg/L based on the linear (slope) of the dose−response curve. This value was between the NOEC and LOEC values. The C_{var} was determined at 5 µg/L, which is near the CC_{20} value. This shows that the C_{var} value approximates the CC_{20} value when the concentration−effect relationship is linear. In Figure 13.3B, where no effects were observed, the calculated C_{var} was 13.5 µg/L and was superior to the applied mean concentrations, suggesting no effects in the

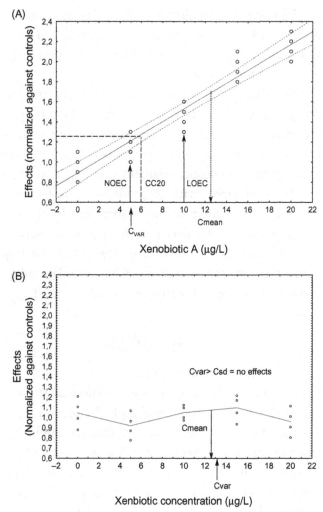

Figure 13.3 Representative linear dose—response trends for toxic and nontoxic compounds. Toxic linear dose—response (A) and nontoxic (B) dose—response relationships are shown. When the sample is not toxic, only the biological variation of the response is apparent. C_{mean} and C_{var} represent the mean and the calculated effect concentrations based on the method of variation.

range of concentrations used. Similarly, the ratio of C_{mean}/k (46 µg/L) was superior to the highest concentration tested (20 µg/L) indicating that no toxicity was observed between 5 and 20 µg/L. Hence, $\underline{C}_{var}$ values > the mean concentrations or the ratio of C_{mean}/k > the highest concentration tested (20 µg/L) indicates absence of effects. We repeated this analysis with semilogarithmic and

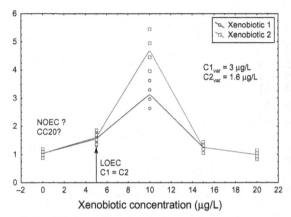

Figure 13.4 Toxicity analysis of U-shaped concentration–response curve.

logarithmic distribution of the xenobiotic concentrations and obtained similar results.

In Figure 13.4, we reported concentration effects following a U-shaped curve for two xenobiotics. In both cases, the LOEC values were the same (10 µg/L) but the mean intensity of the response was 3.1 and 4.7, respectively. Thus, xenobiotic 1 is considered less toxic than xenobiotic 2 because it produced lower effects at the same LOEC concentration. In this case, the LOEC or NOEC values were not suitable descriptors of toxicity. Moreover, the EC_{20} values are not available since the curves are not linear and data points in the linear portion (i.e., between 4 and 10 µg/L) are lacking. The C_{var} values for xenobiotic 1 and 2 were 3 and 1.6 µg/L, respectively. Xenobiotic 2 is more toxic since less concentration was needed to induce variation in the biological responses compared to xenobiotic 1.

With U-shaped concentration–response curves, the evaluation of CC_{20} based on linear regression and even the NOEC values could be difficult to determine. The C_{var} for substances C1 and C2 were 3 and 1.6 µg/L, showing that the xenobiotic C2 was more toxic than the xenobiotic C1.

Another example of a nonlinear curve is shown where significant changes are detected only at the highest concentration tested. In this case, substances 1 and 2 share the same LOEC values but differ in intensity, 1.6 and 2.6-fold responses, respectively (Figure 13.5). Moreover, the EC_{20} cannot be estimated since only one point shows significant differences from the controls and the other concentrations. However, substance 2 could be considered more toxic

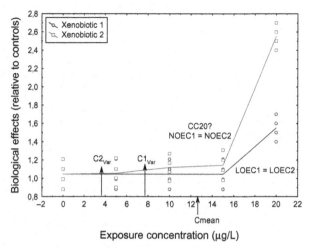

Figure 13.5 Other examples of nonlinear concentration–effects relationships. A concentration–effect is presented where the toxic changes are observed at the highest exposure concentration. The data represent the mean with the SD.

than substance 1 since it produced a more intense response albeit at the same NOEC and LOEC values. This is another example where threshold effects and EC_{20} are not suitable descriptors of toxicity. The calculated C_{var} for xenobiotic 1 and 2 was 7.7 and 3.7 µg/L, respectively, indicating that xenobiotic 2 was more toxic than xenobiotic 1. In this case, C_{var} could be considered a "threshold" concentration that precedes the manifestation of toxic effects and that takes into account the changes in variation of the biomarker response. We tested this with semilogarithmic and logarithmic concentration gradients and found similar results. The C_{var} value was similar to the C_{mean} for the semilogarithmic plot indicating that when changes in responses are below twofold the controls for last concentration used, the toxicity might be considered marginal. Thus, the sample could be considered not toxic when low intensity in the effects (<2) is observed at the highest concentration tested when using semilogarithmic concentration increments.

As a final example, a complex hormetic exposure–effect response is shown in Figure 13.6. Hormesis is found in many instances in toxicological studies and is usually found when dealing with complex mixtures such as municipal and industrial effluents or with highly regulatory physiological systems such as the immune response or endocrine signaling [4,5]. A significant increase in the biomarker is observed at low concentration, which is followed by a gradual

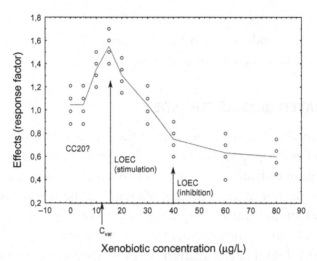

Figure 13.6 Hormetic dose–response curve. In this effect–concentration response curve, a hormetic effect is observed with a significant rise of the measured effect at 15 μg/L with a significant decrease at 40 μg/L, i.e., two LOECs are found. However, the CC_{20} is difficult to determine since the change in effects is not linear. The concentration threshold by which a biological variation is observed was found at 12 μg/L.

decrease response in the effects at higher concentration. In this cases, two LOECs (and NOECs) are obtained, one for the increase and the other for the decrease in the effects: 15 and 40 μg/L for the increased and decreased response, respectively. The CC_{20} based on linear regression models cannot be calculated in this case because the relationship is nonlinear. These types of responses cause the investigator to consider that biomarker responses are dual (non-monotonic) in nature, which makes the investigator consider which response is the most meaningful. The C_{var} value obtained for this type of response is calculated at 12 μg/L or the concentration to reach by which a variation in the biological effect is observed. In this situation, C_{var} value makes no assumption on the type of relationship between the effect and exposure concentration.

In conclusion, a new descriptor of toxicity is proposed that is based on the principle of variation in which the concentration that produces a biological variation in the response (C_{var}) is calculated. The C_{var} is considered the concentration to be reached that induces variations in the biomarker (effect) response. In a simple linear exposure–effects relationships, the C_{var} value is closer to the NOEC than the CC_{20} value. C_{var} values are dependent on both the threshold effects and the intensity of the observed effects but are independent of linear or other monotonic approximations.

13.2 INTEGRATED BIOMARKER INDEX

Studies on biomarkers of effects determine the physiological status of organisms under stress by anthropogenic factors such as pollution, climate change, and loss/transformation of habitats in aquatic ecosystems. Hence, sites that represent a threat to the survival of endemic populations can be determined and identified. Biomarker studies usually require a battery of biomarkers in addition to tissue-specific chemical measurements such as metallothioneins (for heavy metal exposure), lipid peroxidation (LPO; oxidative-mediated damage), energy expenses (decreased energy reserves), and xenobiotic biotransformation of organic compounds. Taken alone they can provide information on potentially deleterious effects and determine the health status of organisms [6]. Integrating a set of biomarkers into an index-based indicator of health has found some merit as an indicator of environmental stress and appears to be a useful tool for environmental managers and ecological risk assessments. There exist many ways to integrate biomarkers such as the star plots of biomarkers [7], the scale of classification method [8], rough set analysis of scaled biomarkers [9], and rank-based biomarker index [10]. Some indexes could be applied in monitoring-type investigations and used as a way to integrate complex data responses into one global index. These approaches could be criticized in some circumstances, but they are useful in large cumulative effects studies where integration is needed to sum up spatial or temporal effects from large data matrices. The scale of classification and ranked-based method will be described next as they are relatively simple to apply, but this is not indicative of the preference of index methodology. All of these approaches have their merits. For example, the star plot, so-called the Integrated Biomarker Response index approach, is gaining more and more attention in European countries [11].

13.2.1 Scale of Classification Method

A classification approach based on the distribution of biochemical markers has been proposed by Narbonne et al. (2005) [8], which makes use of the discriminant factor. Biomarker values are scaled by the discriminant factor, which is defined by the difference between the maximum and minimum value of the biomarker for all the dataset/95% CIs. The discriminant factor sets out the maximum number of classes a biomarker response could have. For example, a fictional biomarker has a Max − Min = 15 and the CI of the dataset is 5. The discriminant factor is then (15)/5 = 3. This indicates that the biomarker value could have 3 classes: biomarker values between 1 and 6 (=min + 95% CI) are assigned class 1, values between 6 and 11 (=min + 2 × 95% CI) are assigned class 2, and so forth. In the original methodology, a different discriminant factor is used that is calculated by [(Max − Min) + 95% CI]/95% CI, which gives extra classes. In the previous example, the discriminant factor would be 4 instead of 3. Class attribution is simply obtained by dividing the Max − Min difference by 4. This approach is then used for each type of measured biomarkers and the class numbers (corresponding to the biomarkers being measured at a given site or treatment group) are then summed up giving an integrated biomarker index. The advantage of this method of scaling biomarkers is that it takes into account the variation for each biomarkers could take at each site and considers the biological variability of the underlying physiological targets.

An example is provided in Table 13.1 which shows the measurement of four biomarkers: metallothioneins, MT; condition factor, CF; DNA strand breaks for genotoxicity; and LPO over three sites where site 3 is considered the least impacted site (reference). At each site, the mean with 95% CI or the SE are given. In the complete data that encompass all three sites for the MT biomarker, for example, the minimum and maximum values are extracted along with the 95% CI. A discriminant factor is calculated as follows: (Max − Min)/95% CI, which gives a value of 6.4. This indicates that the MT values could be scaled at least 1−6 because the 95% confidence limits could be placed six times in the dataset. For example, starting with a minimum value of 0.19 for MT, adding the 95% CI (0.63) gives 0.82. Biomarker values within 0.19 and 0.82 are considered the same and correspond to class 1. Class 2 biomarkers are those within 0.82 and 0.82 + 0.63 = 1.45, and so forth. The mean MT values for each site are thus scaled as follows: site 1, 2, and 3 are 3, 4, and 1. After transforming the remaining biomarkers in a similar manner, an index is produced by simply adding the scaled biomarkers at each site to give 11, 16, and 6 for sites 1, 2, and 3, respectively. Site 2 is clearly more impacted than site 1 and 3 and site 3 is the least impacted in keeping with the reference site.

Table 13.1 Biomarker Scaling According to the Discriminant Function

Sites	Site 1	Site 2	Site 3	All groups
MTs				
Mean ± 95% CI	1.8 ± 0.39	2.22 ± 1.46	0.4 ± 0.12	
Max − Min				4.24−0.19
95% CI				0.63
DF				6
Class	3	4	1	
CF				
Mean ± 95% CI	O.12 ± 0.026	0.13 ± 0.017	0.08 ± 0.044	
Max − Min				0.196−0.044
95% CI				0.022
DF				7
Class	4	4	2	
DNA strands (genotoxicity)				
Mean ± 95% CI	555 ± 321	1258 ± 531	297 ± 81	
Max − Min				165−2694
95% CI				313
DF				8
Class	2	4	1	
LPO				
Mean ± 95% CI	0.259 ± 0.25	0.96 ± 0.4	0.42 ± 0.1	
Max − Min				0.01−1.83
95% CI				0.24
DF				7
Class	2	4	2	
Sum of classes	**11**	**16**	**6**	
(Scale classification)				

DF, discriminant factor.

13.2.2 Rank-Based Approach

Another way to produce a biomarker index is to rank the biomarkers according to the statistical difference between the biomarker responses at each site. Rank-based scaling is similar to the previously discussed approach, but the biomarkers are ranked (scaled) based on the number of sites under study instead of the discriminant factor approach. This approach has the additional advantage of working with nonparametric data distribution. In this procedure, the mean or median value of each site is ranked from $1 =$ lowest value and the number of sites in the study where the highest value corresponds to the total number of sites. Ranks are full integers 1, 2, 3, ... n sites or treatment groups.

Table 13.2 Biomarker Integration Using the Rank-Based Approach

Sites	Site 1	Site 2	Site 3
MTs			
Mean ± 95% CI	1.8 ± 0.39	2.22 ± 1.46	0.4 ± 0.12
Rank	2	3	1
Rank after multicomparison test	2	2	1
CF			
Mean ± 95% CI	0.12 ± 0.026	0.13 ± 0.017	0.08 ± 0.044
Rank	2	3	1
Rank after multicomparison test	2	2	1
DNA strands (genotoxicity)			
Mean ± 95% CI	555 ± 321	1258 ± 531	297 ± 81
Rank	2	3	1
Rank after multicomparison test	2	3	1
LPO			
Mean ± 95% CI	0.26 ± 0.25	0.96 ± 0.4	0.42 ± 0.1
Rank	1	3	2
Rank after multicomparison test	1	3	2
Sum of rank	**7**	**12**	**5**
(Integrated biomarker index)			

A nonparametric multicomparison test (Mann—Whitney U test) is then used to compare between each rank (sites) value. If the difference between rank 1 and 2 is not significant then the same rank number is attributed to each site (rank 1 for both site 1 and 2). Conversely, if the difference is significant then highest mean value is attributed a higher rank value (rank 1 for site 1 and rank 2 for site 2). The process is then repeated between site 2 and 3 and so forth. If there is no difference between sites 2 and 3 then they share the same rank values (rank 2 for site 2 and 3). If a significant difference between site 2 and 3 occurs, then site 3 is ranked at 3 and site 2 ranked at 2. The process is repeated for all sites for each biomarker. The ranked values for each biomarker at each site are summed into an index.

An example is provided in Table 13.2. For the MT biomarker, ranks are first attributed from the lowest (rank 1) to the highest mean value (rank 3 because we have 3 sites). Then rank 1 from site 3 is compared with rank 2 of site 1 using the Mann—Whitney U test. The difference is significant so that rank 2 is attributed to site 1. Rank 2 of site 1 is then compared to rank 3 of site 2. The sites are significantly different thus, site 2 is attributed rank 3. The procedure is repeated for each biomarker. The rank value of the biomarker at each site is then summed where site 3 is the least impacted with a ranked sum value of 5 and site 2 is the most affected with a sum rank value of 12. In this approach, the biomarkers are scaled according to the rank values from each site where higher rank values are attributed based on the statistical significance from the previous rank (site). This rank-based approach could handle nonparametric data distributions where Mann—Whitney U tests are used for rank attribution. The biomarker indexes are summed instead of providing the mean value. This is a reflection of the notion that organisms cumulate various stresses instead of sensing an average value of effects in the real world. In other words it is the accumulation of responses that leads to pathophysiological state, not the average intensity of the effects.

13.2.3 Finding the Normal or Natural Range of Biomarkers

In some occasions, field-based investigations have raised the problem of finding suitable reference sites when determining whether a change in the biomarker response is problematic to an organism's health. In many instances, reference sites are difficult to find because of the plurality of pollutant inputs in populated areas. For example, one investigator wishes to study the effects of a municipal effluent or a marina in a river and samples invertebrates upstream and downstream the area. The upstream site is supposed to be devoid of any pollution or is the reflection of the basal effects of the river system. The upstream site could contain other upstream sources of pollution (e.g., agricultural fields or leaking septic tanks from a village), which can produce a different response pattern. A way to replace a reference site (but still include upstream and downstream sites from a source of pollution) is to understand the normal variation of the biomarkers used in the test battery. This could be achieved by measuring the natural variation of the biomarker(s) over many months in an attempt to understand the normal range in variation. If no clear reference or low-contaminated site could be found, the organisms could be held in aquaculture if possible and the normal range of biomarker responses could be determined. An example is provided in Figure 13.7 where a biomarker is measured for 6 months during spring and autumn. The mean value of a clean and natural site was also measured during that time and the data are reported as the mean (solid line) with 95% CI (dotted lines) in Figure 13.7A or as the median (solid line) with 25th—75th percentile (dotted lines) in Figure 13.7B. The selection of these intervals will depend on

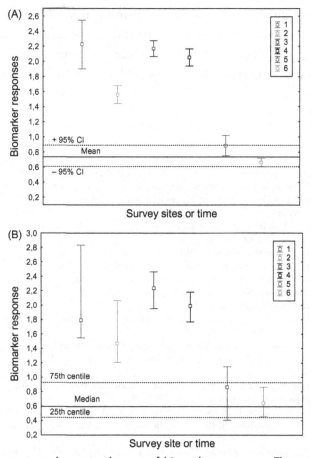

Figure 13.7 Setting a normal or natural range of biomarker responses. The natural range of biomarker response is presented for parametric (A) and non-parametric (B) data.

the study design such as the behavior of the data distribution and replication during each analysis time. The idea behind this approach consists of setting an interval of the biomarker responses that represent the normal interindividual responses in time. Values outside this interval are considered "extreme" or abnormal responses and could represent a threat to the organisms. Biomarker index values could be used as well where the median-based interval should be used. In the example provided in Figure 13.7, samples 5 and 6 display a response that is within the normal range of responses and should be of no concern. However, in sites 1 to 4, the data are expressed outside the normal range of responses in both cases where the normal range is expressed as the mean (with 95% CI) and as the median (with 1st and 3rd quartile). These sites should be of concern for the animal and merit further investigation. This approach is similar to clinical biochemistry for blood or urine chemistry values and has the advantage of alleviating the need for reference sites in a study design. However, it requires defining the interval of safeness with a good number of organisms in space and time.

REFERENCES

[1] Calabrese EJ. Paradigm lost, paradigm found: the re-emergence of hormesis as a fundamental dose response model in the toxicological sciences. Environ Poll 2005;138:379–411.

[2] Calabrese EJ, Blain RB. The hormesis database: the occurrence of hormetic dose responses in the toxicological literature. Regul Toxicol Pharmacol 2011;61:73–81.

[3] Fagin D. The learning curve. Nature 2012;490:462–5.

[4] Prehn RT, Berd D. Whipsaw cancer treatments: the role of hormesis in endocrine and immune therapies. Semin Oncol 2006;33:708–10.

[5] Ren H, Shen J, Tomiyama-Miyaji C, Watanabe M, Kainuma E, Inoue M, et al. Augmentation of innate immunity by low-dose irradiation. Cell Immunol 2006;244:50–6.

[6] Gagné F, Blaise C. Review of biomarkers and new techniques for in situ aquatic studies with bivalves. In: Thompson KC, Wadhia K, Loibner AP, editors. Environmental toxicity testing. Oxford, UK: Blackwell Publishing; 2005.

[7] Beliaeff B, Burgeot T. Integrated biomarker response: a useful tool for ecological risk assessment. Environ Toxicol Chem 2003;21:1316–22.

[8] Narbonne J-F, Daubeze M, Clérandeau C, Garrigues P. Scale classification based on the biochemical markers in mussels: application to pollution monitoring in European coasts. Biomarkers 2005;10:58–71.

[9] Chèvre N, Gagné F, Blaise C. Development of a biomarker-based index for assessing the ecotoxic potential of aquatic sites. Biomarkers 2003;8:287–98.

[10] Blaise C, Gagné F, Pellerin J, Hansen P-D, Trottier S. Molluscan shellfish biomarker study of the Quebec, Canada, Saguenay fjord with the softshell clam Mya arenaria. Environ Toxicol 2002;17:170–86.

[11] Marigómez I, Garmendia L, Soto M, Orbea A, Izagirre U, Cajaraville MP. Marine ecosystem health status assessment through integrative biomarker indices: a comparative study after the Prestige oil spill "Mussel Watch." Ecotoxicology 2013;22:486–505.

Note: Page numbers followed by "*f*" and "*t*" refer to figures and tables, respectively.

Printed in the United States
By Bookmasters